50

50

FIFTY UNDER FIFTY

INNOVATORS OF THE 21ST CENTURY

BY BEVERLY RUSSELL, EVA L. MADDOX & FAROOQ AMEEN

Published in Australia in 2015 by
The Images Publishing Group Pty Ltd
ABN 89 059 734 431
6 Bastow Place, Mulgrave, Victoria 3170, Australia
Tel: +61 3 9561 5544 Fax: +61 3 9561 4860
books@imagespublishing.com
www.imagespublishing.com

The Images Publishing Group Reference Number: 1122

National Library of Australia Cataloguing-in-Publication entry:

Title: Fifty Under Fifty: Innovators of the 21st Century / edited by Beverly Russell, Eva L. Maddox & Farooq Ameen.
ISBN: 9781864706000 (hardback)
Subjects: Designers.
Design.
Architects.
Architecture—Designs and plans.
Architecture—International cooperation.
Creative ability.

Dewey Number: 729

Concept design: Michael Gericke and Matt McInerney, Pentagram
Production manager: Rod Gilbert, Images Publishing
Visual interpretation design: Ryan Marshall, Images Publishing
Senior editor: Gina Tsarouhas, Images Publishing
Proofreader: Jade de Souza, Images Publishing

Pre-publishing and printing by Everbest Printing Co. Ltd., in Hong Kong/China, on 140gsm GoldEast Matt Art paper

IMAGES has included on its website a page for special notices in relation to this and our other publications.
Please visit www.imagespublishing.com.

Disclaimer: Any additions / corrections to this book will be made by the authors in future editions.

Contents

Overview

Beverly Russell

After months of careful scrutiny and research by a bevy of informed people, with the final selection made by a distinguished five-person jury, namely Jeanne Gang, Ralph Johnson, Qingyun Ma, Stanley Tigerman and Marion Weiss (see page 328), we present here a troop of men and women all under the age of 50 in 2015; 50 design and architecture firms across the world. All have a singular aim: to express relevant architecture for their time—to design out of the box.

As an international group spanning from Shanghai to Los Angeles, Amsterdam to New Delhi, and presenting work across 19 countries on five continents, including the North and South Americas, Europe, Africa, and Asia, these visionaries are holistically conscious of global needs, and they perceive architecture and building through the lens of humanitarian structures, acutely aware, in the words of prophetic writer H.G. Wells, that "civilization is a race between education and catastrophe."

Born during the latter part of the 20th century, this is a generation familiar with adaptation. They have waded through drastic political, economic and social upheavals with academic analysis, unlike the younger millennials (aged 18 to 32) who seemingly have absorbed cyberspace acrobatics by osmosis.

The firms featured here exhibit passion for the planet, concern for the future, and joyful satisfaction in creating spaces that respond to humanistic realities: schools, medical clinics, integrated agriculture opportunities, sustainable low-cost housing, affordable effective transportation. Clients with deep pockets also lend these designers the opportunities to further render into reality proposals for intelligent thinking in the spirit of traditional shelter, work, and entertainment spaces.

But the 50 standouts realize the urgency to reinvent mechanical systems, to readdress building techniques driven by an expanding global population of almost 7.5 billion people, one third of whom live in poverty.

Even in China, where mega starchitects compete to design ever-more bizarre skyscrapers, there is a concern that the focus on big may not be better for the population. China, along with India, South America and Africa all count millions of people with barely a roof over their heads or who cannot read or write. Moving these populations to cities often results in chaotic urban slums, totally counterproductive to the mission to generate a better quality of life. Some of the solutions are to be found in retaining rural communities with dignity. This is the breeding ground for innovative ideas that are sweeping through the architectural profession today, as Farooq Ameen illustrates in his observations on current design leadership in Relocating Place, Form and Identity (see page 21)—citing 2014 Pritzker Prize-winner Shigeru Ban, and the influential Dean of Architecture at the University of Southern California, Qingyun Ma.

These architects and designers speak of a new balance, observing that traditional, static cities have become more environmentally dynamic. They ask us to imagine water-based floating architecture that can offer an answer to climate change—critical for flood-endangered locations. They appeal for urbaneering, that is, a restructuring of metropolitan housing that is interwoven, perhaps with elevated platforms for food production.

In selecting these innovators one aspect became crystal clear. Geographical origin transcends the customary formula for the work. Beauty cannot be a north star to guide opinion on the selection process. As juror Qingyun Ma affirmed: "To judge only from the point of view of visual quality would be unfair." He noted that many architects and designers trying to find adequate solutions for under-served developing communities must use the materials available and the labor force at hand for construction. In their final form, these buildings cannot in any way compete with Western aesthetics and yet they exhibit innovations that yield an alternative benchmark for excellence. They incorporate unique local intelligence, thus informing a new set of design rules that must be respected for their geographical originality. A recent Venice Biennale with its theme Common Ground—focusing on urban agriculture and nature—sought to investigate "cities in the countryside" seeking an alternative to skyscraper metropolises shrouded in pollution with diminishing benefits for human activities.

It is just over 20 years since Stanley Tigerman, chair of the jury for this book (see page 346), launched Archeworks in his home city of Chicago, an alternative school devoted to finding design solutions for marginalized communities. Eva L. Maddox, this book's co-author (see Innovation 2015, page 14), was his partner in this pioneering enterprise, which is ongoing today. The school enrolls students from all over the world, to facilitate the filtration of Archeworks' message beyond United States' borders. This brave stab at conjoining architectural education with a dedicated commitment to social responsibility was initially greeted with some doubt by professionals, at first. However, Archeworks has consistently fulfilled what philosopher-writer Thomas Merton calls "the invisible fecundity" in its mission, "based on value, rightness and truth." The school has helped minority constituents in Chicago to turn a page in their existence. It has educated women to use their own skills to move from welfare to workfare; it has provided alternatives for the homeless and breakthrough inventions for the physically and mentally challenged; and street improvements to resurrect viable urban spaces.

While Chicago has an important history in the art of building, California on the west coast of the United States has also captured pioneers and creative thinkers. Following the gold rush (which forecast a transient future of instant wealth through ingenious engineering) came the oil boom. Historians have cited this era as one of audacity and good luck—a new American dream compared to the dreary puritanical mien observed in other regions of the country at the time. Early in the 20th century the west coast began to draw another group of pioneers, photographers from Europe. They found magic in its wide, open spaces, benign climate and luminous quality of light, birthing (among many other industries) the art of movie-making. They helped build the celluloid city of Hollywood—Tinsel Town—and founded Los Angeles as the center of world entertainment.

Along with the movie-makers came the architects and designers, to build the lots, the sets, the offices and the residences to accommodate this new booming industry. Architects such as Frank Lloyd Wright, John Lautner and Richard Neutra were propelled to new paradigms, to design houses that matched a sunlit indoor-outdoor lifestyle built on the hills and canyons of this rugged terrain. In the 1940s and 50s the Case Study Houses attracted international acclaim for their new modern approach to materials and structure. Architectural innovation, born in California, continues to make headlines across the globe, with Frank Gehry's renowned buildings in Spain, France, the UAE, and Australia, and the work of another Pritzker Prize–winner, Thom Mayne, whose international projects are gaining momentum.

Today, Silicon Valley in northern California and Silicon Beach in the south are the nerve centers of extraordinary digital invention. Apple, Facebook, Google, Microsoft, Twitter, Instagram, and others, lead a pack of industrial designers and mathematical engineers who have invited us into the virtual experiences of the 21st century.

It should be no surprise, then, that 16 firms (30 percent) of the visionary 50 published in this book are based in California while working, very often, for global clients. Many of these partnerships comprise collaborators from other countries. These designers find abundant creative energy in this region saturated with the collective consciousness of frontrunners in all the arts. They are the next generation, integrating the highest tech developments from NASA's Jet Propulsion Laboratory, inventors of the Curiosity rover on Mars, and the Silicon internet wizards whose creativity ripples around the globe. They have reinvented the *California Dreamin'* image.

Clockwise from top right: Hip Hop celebrity Cesar Rivas on his head in one of the dance studios at the Hip Hop School of Arts; the youth-appealing logo for the school; eye-grabbing mural painting by Hector Rios; Ryan Lopez, project architect for HHSA located in a remodeled abandoned bank building; students in session at the school.

Photography by Barry Michlin

One typical example of this innovative mindset is the Hip Hop School of Arts, founded and funded by celebrity hip hop artist Cesar Rivas in Pomona, California, a marginalized city of 152,000, which illustrates typical United States urban demographics: 44,000 under age 18; high-school drop-out rate 47 percent; crime rate 26 percent; obesity 29 percent. The school's mission (totally funded by private donors) offers what Rivas calls a "way out," an opportunity to be educated in a trade (within the hip hop culture) that will sustain a career; tuition is free for qualified low-income families. The school is housed in a formerly abandoned main street bank building, now remodeled into a thriving educational mecca by 36-year-old Woodbury University architectural graduate Ryan Lopez, a project architect in the office of Pasadena architect and garden designer Virginia Paca. Students learn a strong work ethic to be able to carry a skill into their chosen profession. The state-of-the-art facility has a recording studio, three dance studios, art room, DJ room, computer lab, green room, and hip hop library.

With gender equality around the world on the front burner, it is noted that 15 of the 66 individuals in this book (25 percent) are women, and this female design contingent is playing catch-up fast and is set to keep on growing. Furthermore, this number helps to demonstrate a universal message in terms of emancipation in the design and architecture fields across the globe: these women are from China, Mexico, Hong Kong, Turkey, India, Iceland, Norway, and the United States, with the United States (five) just edging out China (four) by one woman. In light of the three-woman American indie rock band Sleater-Kinney singing hit records in support of women's rights to its global YouTube audience (followers number in the tens of thousands), it's apt to quote the gurus in Silicon Valley: "Failure is not an option."

Innovation 2015

Eva L. Maddox

In 2015, design has found its way. The architects, designers, artists and others represented in *Fifty Under Fifty* are innovators of our time, offering in their work and actions a passion for life and humanity. They have shared with us, and the world, their desire and excitement for exponential learning. It is their individual curiosities, imaginations and talents that allow them to bridge disciplines, respect cultural norms, respond to human needs regardless of costs, harvest their craft, and adopt team transparency in their passion to create and solve problems with a clear mission. These new dimensions thrust them into the forefront of global design innovation. And never before has design innovation itself been at the forefront in a way so exposed, anticipated, and expected.

A remarkable shift is occurring in the global geography of innovation as the world becomes more interconnected through technology. For the past 50 years "objects" of innovation have been created in an incremental, linear mode, an improvement upon improvement based on need, discovery or often value.

These design innovations have formed the basis for our expectations and have become the visual models for idea advancement worldwide. Today, with transparency and interconnectivity, the democratization of design has exposed and exploded innovation around the world. Design innovation, similar to the scientific discovery of our DNA, now has open access, allowing it to take many different nonlinear paths with unexpected consequences.

Apertures, Baumgartner+Uriu, "The project emphasizes an architecture between nature and technology, which can operate as an interactive building organism where multiple discrete features work simultaneously and independently. In this case, sound is used to bridge the gap between the natural and the artificial, allowing the visitor to experience their own biorhythms."

Photography by Josh White

What is design innovation in 2015? What are the new metrics to credit good design? How can a jury look across the world, respond to cultures, understand values foreign to their understanding and economic resources? The *Fifty Under Fifty* innovators possess several common characteristics:

Purposeful value

Architecture for design's sake is no longer the sole purpose of their talents, and does not satisfy their innate desire to add value to society through their work.

Team contributions

They are redefining innovation and design to include "heart" and the notion to "do good." Additionally they have given the world access to their design processes and design thinking, making their design knowledge a human right for all.

Multidiscipline achievements

Many of these innovators are sharing their talent and work by joining forces with other industries in order to achieve the unthinkable. Just as medicine, science, technology, and other professions, are teaming to accelerate innovation, architects and designers are also looking beyond their own profession to achieve remarkable discoveries.

Portable House ÁPH80, ÁBATON, "The solid timber structure has been fully designed and manufactured in Spain through numerical control. Manufacturing time takes eight weeks. Assembly time takes one day and transportation is by road. Currently, we are developing a simpler series, which can be added to the ÁPH80 to suit every need, creating larger spaces and contributing to the project's versatility."

Photography by Juan Baraja

Kwe Ka Baung School, a.gor.a architects, "Students, as well as those involved in the two-phase construction process, will benefit from this project. Several Karen workers are refining their skills in adobe brick-making, while contributing to the design that we implemented. The program aims to show the community a feasible way to pursue affordable architecture without compromising spatial quality, while maintaining a strong connection to the landscape."

Photography by Abel Echeverria

Defence Colony Residence, vir.mueller architects, "As the new Indian architectural culture seeks international expression, this project expresses the vitality and beauty of building with a local, sustainable and basic material. It also sustains the livelihood of a guild of brick masons struggling to retain their place in society."

Photography by André J. Fanthome

Chengdu Fluid Core-Yard,
P-A-T-T-E-R-N-S

Photography by P-A-T-T-E-R-N-S

Samsung Raemian Master Plan,
Contemporary Architecture Practice

Photography by Nacasa and Partners

Lost Luggage: mixed 3D printing technologies,
Janne Kyttanen BV

Photography by Adrian Woods & Gidi van Maarseveen
© 3D Systems

Advanced materials and making

"The *Fifty Under Fifty* are among an emerging generation of architects and designers, conducting a robust body of research which lights the way to new materials for building and making."[1] No longer does a material serve just one leg of the stool. Society is beginning to see the evolution and the value of organic, living materials along with manufactured materials for their design properties, lifecycle and resilience. These designers are embracing the fact that science and biology are fostering a leap to stealth materials, organic experimentation and technology that's accelerating the creation of a new sustainable architecture.

Neri Oxman, Israeli architect and Professor & Research Group Leader at MIT Media Lab, transcends the boundaries between art, science and environmentalism. She is an excellent example in addition to these chosen innovators. Her stated goal is to "enhance the relationship between the built and natural environments by employing design principles inspired by nature and implementing them in the invention of digital design technologies." In defining these undervalued materials, Neri Oxman notes "the story of the material is where its value is determined." This is of particular interest with the 50 chosen innovators as many have identified materials of multiple values and uses. Their exploration includes basic building materials of their culture (i.e. bricks and mortar), natural elements, and living organisms.

Gando School Library, Kéré Architecture, "Casting the ceiling with locally produced earthenware pots creates day-lighting effect."

Photography by Kéré Architecture

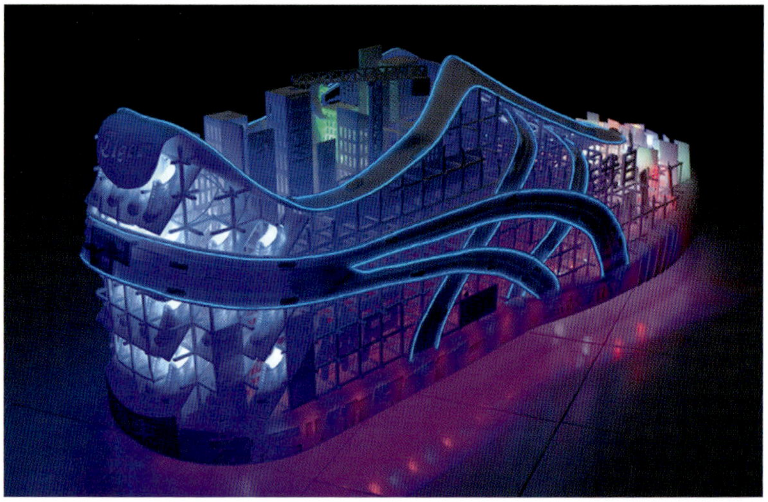

Electric Light Shoe, Janne Kyttanen BV, "The Electric Light Shoe was created for the OnitsukaTiger 'Electric Tiger Land' campaign. These images show the 1-meter-long (3-foot) Electric Light Shoe with illumination. This design was central in the brand campaign and features a miniature microcosm that mimics the hustle and bustle of Japan's urban centres."

Photography by Amsterdam Worldwide

Global sharing and no boundaries

The 50 selected architects and designers exhibit attitudes of sharing, caring and giving back. They are out to save the world, the best way they know how—by pushing boundaries. As they stretch their imaginations, share their desires and spark worldwide design dialogues, they will lose no time in finding alternative solutions to the issues of the day. Their talents are unique. Their energies are contagious. They use additional skills and senses to achieve their goals, resulting in designs that are alive, transparent, untethered, and simply beautiful.

Working differently

This new breed of innovators works differently. They "come together and connect in special districts, rather than working alone in isolated spaces. Special districts can mean many different types of places for innovation, such as the internet, a space within the community, in urban districts, in educational institutions, in start-up campuses and accelerator labs."[2]

Spurring innovation and economic development

The *Fifty Under Fifty* innovators have the unique potential to spur productive, inclusive and sustainable economic development—bringing people, process and place together. No longer are the mad scientist and the artist architect working alone. They are working hand-in-hand to spark ideas, produce new discoveries, expand employment opportunities for the disadvantaged, create shelter for catastrophic events, save the environment for future generations, and explore new materials and uses for sustainability—all with respect to cultural heritage.

Economic assets are evidenced by the innovator's day-to-day risk-taking culture and involvement, responding to social need to always create and surprise with new answers/solutions. They work with "innovation drivers, innovation cultivators and neighborhood building amenities" to achieve innovative goals.[3]

A broad brush of the *Fifty Under Fifty* innovators expands our understanding of the interpersonal connections, cultural ties, and global perspectives that these talents build through networking, including (or as evidenced by their) community outreach and public art, aquatic learning platforms, purposeful educational furniture, and fun-loving "kitten" desks, government buildings in India, architectural jewel outposts in Norway, useful commercial objects, and exhibitions to challenge the status quo. They choose to "address many issues in their problem-solving with an understanding of simplicity and beauty."[4]

The 50 under 50 showcased in this book are a network of pings bouncing back and forth, continent to continent, leaving their spirit and idea markers to accelerate new ways.

1 Lara Kristin Herndow and Derrick Mead, "7 Material Innovations That Could Change Design," *Metropolis* magazine.

2 Bruce Katz and Julie Wagner, *The Rise of Innovation Districts: A New Geography of Innovation in America*, Brookings Institute, December 29, 2014.

3 Ibid.

4 Ibid.

Wetslums, Koen Olthuis, "One billion people live in slums and 30 percent of them in wetslums. By leasing floating functions in containers to entrepreneurs in slums, the life conditions can be upgraded. These so-called City Apps are flood-resistant, take only available space on water and can be moved to other locations and slums when situations change."

Rendering by Waterstudio.NL

Makoko Floating School, NLÉ Works, "NLÉ has developed a prototype floating structure that will serve primarily as a school, whilst being scalable and adaptable for other uses, such as a community hub, health clinic, market, entertainment center or housing. The prototype's versatile structure is a safe and economical floating triangular frame that allows flexibility for customization and completion based on specific needs and capacities."

Photography by NLÉ

Relocating Place, Form and Identity

Farooq Ameen

Leading Innovation Today

Fumihiko Maki, the remarkably modest Pritzker recipient and architect of the new Four World Trade Center complex in New York, recently discussed his manifesto to a spellbound audience of nearly 3000 individuals in the most dense city in the world: Dhaka, Bangladesh. It was not a coincidence that some of the world's most creative designers, historians and activists had descended upon this unbridled South Asian megalopolis. The challenges facing its future and that of the large majority of cities will demand the most innovative design solutions. Maki's ephemeral work is grounded by the timeless Vitruvial essentials of *Utilitas*, *Firmitas*, *Venustas*, (Durability, Utility, Beauty), of which the third is the most critical for him. The real test, according to Maki, is not visual tectonics but that the users experience delight (*Venustas*) by claiming the work as their own. This is the true measure of success in design innovation.

The primary proponent of the field of material ecology, Israeli architect Neri Oxman reminds us that the act of imagination is the key to nurturing innovation and perhaps the most critical aspect in a designer's creative agenda. This is certainly true of the select group of designers whose work and philosophy have been presented in this volume. Whereas the imaginations of the 50 remain uniquely their own, they are exposed directly or indirectly to the innovative work of accomplished designers who are leading the innovative edge in all aspects of the design of the environment. We might organize these aspects in several categories to highlight key innovative principles: ecological urbanism, bio-mimicry, reinventing fabrication, rediscovering material, architectural alchemy, hedonistic sustainability, and relocating tradition. As is evident in the work presented here, innovative projects incorporate multiple aspects resulting in a complex, layered response to the challenge of place, form, and identity.

Ecological urbanism: think like a king, act like a peasant

The dynamic Beijing-based landscape architect Kongjian Yu calls his practice Turenscape, derived from two Chinese characters: *Tu-Ren*. He explains that Tu means "land" or "earth" and Ren means "human being" or "man"; and that the ideology of his practice (literally earth-man landscape) is about creating harmony between land and people. Eschewing the ornamental and cosmetic that is geared toward a high-culture-seeking pleasure, he declares that China's landscape, like that of many other rapidly growing Asian nations, is about survival from a crisis of energy and a shortage of clean water. Yet, it is the scale and pace of the challenge that has enabled Turenscape to implement ideas that are generally considered theoretical in the West, according to William Saunders, editor of *Harvard Design Magazine*.

The Shanghai Expo Houtan Park is a 34-acre (13.8-hectare) regenerative landscape built on a former brownfield industrial site on the Huangpu riverfront in the heart of Shanghai. The primary ecological strategies included the construction of a 1.06-mile-long (1.7-kilometer-long) wetland that is a living machine treating contaminated water; it cascades to oxygenate the nutrient-rich water and a series of terraces that enable connectivity in the nearly 16.5-foot (5-meter) separation between the city and the river. A habitat-friendly riprap protects the shoreline and enables native species to grow.

Shanghai 2010 Expo Houtan Park, Shanghai, China, Turenscape

Photograph courtesy of Turenscape

Similar regenerative techniques slow down stormwater flow at the nationally celebrated Liupanshui Minghu Wetland Park, a formerly deteriorating suburban site on a channelized concrete river. In addition to stormwater management, the park cleanses the water, recovers native habitats, and creates dynamic public space. In both these projects, Yu challenges us to consider bold ambitious strategies, but to learn from vernacular techniques, to leave aside the cosmetic (elitist small feet aesthetic) for the regenerative (peasant big foot survival). In his words, we must "think like a king, act like a peasant."

Material ecology: bio-mimicry, computation, and fabrication

In juxtaposition to the vernacular bio-mimicry and grand scale of Turenscape's work in China, Neri Oxman's goal is to "enhance the relationship between the built and the natural environments by employing design principles inspired or engineered by Nature." She suggests that the emerging paradigm is that of the "world-as-organism" where there is a desire to "instill intelligence into objects, buildings and cities." She contrasts this with the industrial revolution, or the "world-as-machine." The field of material ecology is thus grounded in the premise that the "age of biology" will overcome the "age of the machine." At the Mediated Matter design research group at MIT, she directs research in an arena that emulates nature at the confluence of computational design, digital fabrication, materials science, and synthetic biology.

These four fields are brought together in the design and fabrication of the two-part Gemini Acoustic Chaise Lounge, using 3D printing technology. The chaise features an enclosure that cushions the body within a colored, multimaterial, 3D printed cocoon, replicating the tranquility of the womb. A solid wood shell exterior protects the composite digital color lining, thus combining natural and synthetic materials. The technology enables specific pressure points on the body to act as a soundproof anechoic chamber. The Gemini and similar projects replicates nature's ability for multi-functionality and mass customization as opposed to the mass production of the industrial revolution. Oxman declares that when "matter, fabrication and environment are integrated into an undifferentiated scheme ... architectural design will have arrived at an ecology of the artificial: a material ecology".

Kontum Indochine Café, Kontum, Vietnam, Vo Trong Nghia Architects
Photography by Hiroyuki Oki

Nine Bridges Country Club, near Seoul, Korea, Shigeru Ban Architects
Photography by Hiroyuki Hirai

Wind and Water Bar, Phu Tho district, Thu Dau Mot Town, Binh Duong, Vietnam, Vo Trong Nghia Architects

Photography by Hiroyuki Oki, Vo Trong Nghia Architects

Reinventing material and craft: green steel of the 21st century

The inevitable connection between social justice, sustainability and material use is gracefully articulated in the iconic work of Vo Trong Nghia. He observes that "Green architecture ... elevates human life by embracing the powers of the sun, wind and water into living space. If the current way of thinking does not change, sooner or later citizens will actually live in concrete jungles." This holistic attitude informs the work which has developed an architecture derived from embracing local materials and traditional skills with a modernist aesthetic and technology. In fact, Vo believes that bamboo will replace other materials and is the "green steel of the 21st century."

The Wind and Water Bar is an elegant expression of this aesthetic, enclosed in a bamboo structure located in the middle of an artificial lake and using prevailing wind combined with the cool lake water to provide natural ventilation; an open skylight at the top of the bamboo dome functions as an exhaust mechanism for the warm air inside. The structural bamboo arch system for the main frame integrates 48 prefabricated units, each of which is made of multiple bamboo elements bound together and built by local workers over a three-month period. Vo acknowledges learning about the Japanese "way of thinking toward climate and natural features" and discovered similarities with that of Vietnam, where he has incorporated them.

The Pritzker recipient Shigeru Ban also learnt from his native Japanese traditions through carpenters working at his parents' home and was captivated by the tools and construction, wanting to become a carpenter. His work is characterized by an elegant innovation that builds on this resourceful access to traditiwonal methodologies. The Pritzker jury recognized this experimental approach and cited his ability to "see in standard components and common materials, such as paper tubes, packing materials or shipping containers, opportunities to use them in new ways. He is especially known for his structural innovations and the creative use of unconventional materials like bamboo, fabric, paper, and composites of recycled paper fiber and plastics."

Ban originally proposed paper-tube shelters in response to the 1994 conflict in Rwanda. The techniques were further developed after the 1995 earthquake in Kobe, Japan, where he developed the Paper Log House for Vietnamese refugees. The aesthetics of cardboard tubes and bamboo have also been extended to conventional projects, like the Centre Pompidou–Metz or the Nine Bridges Country Club, which have undulating latticework roofs comprised of wooden strips inspired by a woven bamboo hat. Similarly, the Aspen Art Museum, a 33,000-square-foot (3,066-square-meter) structure in Colorado has a woven exterior wood screen. As noted by the Pritzker jury chair, Lord Peter Palumbo, Shigeru Ban's work demonstrates an "emphasis on cutting-edge materials and technology; total curiosity and commitment and endless innovation".

8 House, Copenhagen, Denmark, BIG-Bjarke Ingels Group
Photography by Jens Lindhe

West 57th, Manhattan, New York, United States, BIG-Bjarke Ingels Group
Image by BIG&Glessner

Architectural alchemy and hedonistic sustainability

The innovative Danish architect Bjarke Ingels considers his projects as opportunities for crafting an architectural alchemy where "you take traditional ingredients that would separately be just "normal this" and "normal that," and when you combine them, because of symbiotic relationships, you get much more out of the mix than if you were to leave them separate". The work extends this perspective to a notion of hedonistic sustainability where comfort is creatively integrated, not compromised. He suggests that architects need to expand their role to become "designers of ecosystems" and not just individual buildings with static programs.

These notions are evident in the 8 House in Copenhagen, comprising a hybrid of offices and shops, two-story garden townhouses, and classic apartments. These are blended together in an infinity loop to form a figure-eight-shaped perimeter block where the row houses are located along a mountain path from the ground floor to the penthouse. This allows people to bicycle or walk all the way to the 11th floor and extends the public realm. The apartments are placed at the top, benefitting from sunlight and fresh air, while the commercial program unfolds at the base of the building merging with life on the street. There are two sloping green roofs strategically placed to reduce the urban heat island effect as well as to visually tie it back to the adjacent farmlands. The shape of the building allows for daylighting and natural ventilation for all units, a sort of hedonistic sustainability where form is shaped by environmental response. In addition, rainwater is collected and repurposed through a stormwater management system.

Bjarke Ingels Group's reinvention of the "New York apartment building" is a hybrid between the European perimeter block and a traditional Manhattan high-rise. West 57th has a unique shape that combines the compactness and efficiency of a European courtyard building providing density, intimacy and security, with the expansive views of a skyscraper. Its unique geometry opens up the courtyard to the Hudson River while bringing low western sun deep into the block. The work affirms the philosophy that "by hitting the fertile overlap between pragmatic and utopia, we architects once again find the freedom to change the surface of our planet, to better fit contemporary life forms".

Relocating tradition

Kenneth Frampton's seminal 1983 essay "Prospects for a Critical Regionalism" recognized that its practice "seeks to deconstruct universal modernism in terms of values and images which are locally cultivated, while at the same time, adulterating these autochthonous elements with paradigms drawn from alien resources." The most prominent efforts until recently have been in the so-called developing world exemplified by the work of Geoffrey Bawa in Sri Lanka, Luis Barragan in Mexico, or Balkrishna Doshi in India resisting forces of internationalization. In this millennium, we live in a reverse paradigm where many prominent designers from Asia, Africa and South America are incorporating traditional insights with practices located in Europe or North America.

Smithsonian National Museum of African American History and Culture, Washington, DC, United States, Adjaye Associates

Image courtesy of Adjaye Associates

Fathers House, Xi'an, China, MADA s.p.a.m.

Photograph courtesy of MADA s.p.a.m.

In a recent interview, David Adjaye observed that his African-British identity gives him "an altered perspective from others in the profession, in which Mies van der Rohe or Palladio have their place, but so do mud constructions in Mali." The approach is to draw inspiration from many influences around the world with an enthusiasm for issues of place and identity. For the Smithsonian National Museum of African American History and Culture, the entire building is wrapped in an ornamental bronze lattice as a historical reference to African American craftsmanship. The density of this lattice pattern, which is integral to its identity, can be modulated to control the amount of sunlight and transparency. In strong contrast, at the Moscow School of Management, the built form responds to the harsh cold winters of Skolkovo. The main elements of the building are assembled as a single entity in which a range of facilities is internally connected and made accessible without going outside as on a traditional campus.

The work of Qingyun Ma is based on a "transcultural mission and cross-Pacific practice" that is steeped in a deep respect for Chinese tradition yet informed by a robust global sensibility. This bi-continental practice (MADA s.p.a.m.) and Ma's dual role as a prominent educator in Southern California consolidate the access to some of the most innovative thinking in North America. His strategy to integrate these notions and gain acceptance and community ownership of his work is to involve local craftspeople and builders. He considers the design of his Father's House in Xi'an, China as a journey that is a dialogue with his father.

The modernist vocabulary and details are not traditional, but the material is local and the technology is familiar, and was built by villagers his father grew up with. Eventually, the house has become a temple (Miao—something unfamiliar) symbolizing the Chinese notion of "presence" that connects his father's past with the future.

In a similar approach to repositioning tradition, the Xian Television and Broadcasting Center takes its form literally from the image of the city of Xi'an where the Wall is the definition of the City. In the newly designated tourism development zone, traditional architecture has been adopted as the official style. Ma reconsiders the tradition by deforming the wall, thereby manifesting an evolution and differentiation from the past. This attitude about relocating tradition consolidates Ingels observation that "Architecture seems to be entrenched in two equally unfertile fronts: either naively utopian or petrifyingly pragmatic." He believes that there is a third way wedged in between these diametrical opposites and calls for "a pragmatic utopian architecture that takes on the creation of socially, economically and environmentally perfect places as a practical objective." In an increasingly connected global paradigm, the viability and competitiveness of our communities will depend on our ability to innovate and evolve while relocating our notion of place, form and identity.

Proje

cts

Jan Glasmeier
Principal

Albert Company Olmo
Principal

476 Moo 1, Soi Sahakorn 2
Intarakiri Road
Tha Sai Luad
Mae Sot 63110
Thailand

www.agora-architects.com
agora.architect@gmail.com
+66 880744229

The practice was founded in 2012 by Albert Company Olmo and Jan Glasmeier. Our firm is based in Mae Sot, Thailand, and works with marginalized groups, community based organizations, and NGOs along the Thai-Myanmar border and Chiang Mai.

Besides working on several school projects, we are the consulting architects on the local Mae Tao Clinic, a health center which provides free healthcare for Myanma refugees.

Our goal as architects is to promote sustainability in construction by introducing alternative materials, such as bamboo and adobe, promoting the importance of mapping projects, creating networks among architectural organizations, and conducting workshops for local staff.

rchitects

Kwe Ka Baung School

In the lime-green landscape of Mae Pa, in the Mae Sot district of northern Thailand, beside a round pond and lotus flowers, we built the second phase of the construction of Kwe Ka Baung School between October 2013 and March 2014. The project aims to provide education to displaced children who have arrived with their families from Karen State, in neighboring Myanmar, fleeing the ongoing civil conflict.

Students, as well as those involved in the two-phase construction process, will benefit from this project. Several Karen workers are refining their skills in adobe brick-making, while contributing to the design that we implemented. The program aims to show the community a feasible way to pursue affordable architecture without compromising spatial quality, while maintaining a strong connection to the landscape.

The Canadian aid organization Casira contacted us in March 2012 and asked us to find a project in which they could get involved and for which they could raise funds. After fundraising $25,000, some 10 members of Casira traveled to Thailand to participate in a four-week workshop. In November 2013, the same group returned to Mae Sot to help with the completion of the school complex.

Photography by Ibai Rigbi (below and right top), Abel Echeverria (right middle)

Mae Tao Clinic, Training Center Facility

Mae Tao Clinic is a humanitarian organization based in the Thai town of Mae Sot, a few kilometers from the Burmese border. Besides providing free medical treatment, shelter and dry food for over 3,000 children, Mae Tao clinic also started an education program in healthcare, sanitation, and disease control for Burmese migrants. This aims to develop a social fabric in the border region and inside Myanmar within the foreseeable future. Space at the clinic has been completely dedicated to serving the ever-increasing volume of patients so Mae Tao Clinic had to find new land to build a facility in which to conduct these training programs.

Adobe, as a natural and environmental friendly material, was used for the classroom and office walls. To protect these walls from standing water, a one-foot (30-centimeter) raised concrete slab had to be built first as a base. The roof is a timber construction with composite roof tiles. The windows and doors, also made out of reused timber, were later painted green and blue for the classroom and ochre for the office building. The materials used are locally available and well known for their uses, allowing for easy maintenance and resulting in low costs.

Photography by Franc Pallarès-López

Mae Tao Clinic, Training Center Dormitories

The armed conflict that has persisted for decades in the Karen State[1] of Myanmar results in a daily flow of refugees and immigrants to neighboring Thailand. In the Thai town of Mae Sot, a few kilometers from the Myanmar border, numerous schools and orphanages offer accommodation and education for the refugees and immigrants.

The lack of space and the need for immediate accommodation for new students has forced the school to present a new model of temporary, low-cost dormitories, which are easy to assemble and can be built by using only recycled material. The temporary dormitories and training classroom buildings are part of Mae Tao Clinic's new Training Center Campus. The Embassy of Luxembourg in Bangkok funded the construction of four dormitories, the first of which was built in four weeks in April 2012.

With a capacity of 25 students, the building meets the *modus vivendi* by fitting into its local environment. The interior layout ensures an open and airy space, which offers semi-privacy and includes storage space for students.

Photography by Allyse Pulliam (above top); Franc Pallarès-López (above bottom and right)

1 In 1989, the ruling military government decreed English transliteration changes to many geographical names. Burma became the Republic of the Union of Myanmar, commonly known as Myanmar. Karen State was renamed Kayin State, although the former is still widely used.

Camino Alonso
Partner

Calle Ciudad Real, 28
Madrid
Spain

www.abaton.es
camino@abaton.es
+34 (91) 352 1616

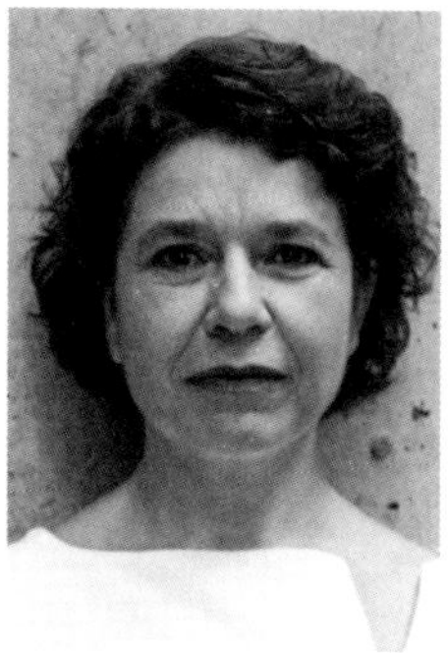

In 1998, Camino Alonso (Architect), Ignacio Lechón (Architect) and Carlos Alonso (Industrial Engineer) founded ÁBATON Arquitectura. Our original objective was to develop housing solutions and architectural projects with a certain degree of freedom. In order to manage the whole building process, we started up a construction company to implement and develop new construction systems in a more flexible way.

From the beginning, Camino Alonso has been the creative manager. She has a special ability to create simple yet human spaces with a unique style. Carlos Alonso is general manager and remains actively engaged in the designing process. Ignacio Lechón represents all ÁBATON projects legally. He also coordinates the construction, urban, and landscape areas.

Currently, we have over 80 completed projects (hotel, office, and restaurant) and more than 50 fully refurbished projects in Spain and internationally.

Camino Alonso began her professional career assisting in the architecture studio of Ricardo Lajaro Olmo. Later, she joined architects Díez y Quero Asociados S.L., engaged in town planning work as well as projects in Alicante, the Balearic Islands, the Canary Islands, Andalusia, and for councils and public authorities.

In 1998, Alonso founded ÁBATON Arquitectura, together with Ignacio Lechón and Carlos Alonso. From the outset, they realized the significance of controlling the entire construction process. They established their own construction firm in order to implement and develop new construction systems and new solutions to projects with greater flexibility.

ÁBATON Arquitectura

Camino Alonso is the second of six brothers and sisters. Although her calling was to be a doctor, her university entrance exam result did not meet the requirements for the faculty of medicine. Instead, she was accepted to study architecture by means of a draw and on her father's recommendation. However, in her fourth year, Alonso returned to the idea of becoming a cardiovascular surgeon but, after the first year, she became convinced that architecture was her true calling. She met Ignacio Lechón, her husband and partner, at university and they have two teenage children.

Since childhood, Alonso has had a long association with the countryside, where she spent extended summer periods (her father came from León). She has never abandoned it and she tries to reproduce it in all of the projects that she takes on. In her home in Madrid, designed by both Alonso and Lechón, she has hens, dogs, and an orchard, looked after by her children. She is a committed woman, with great ability, who has learned how to balance her family and her work successfully.

With this in mind, she strives to enable her colleagues to enjoy a work/life balance as well and her studio is a pioneer in reconciling these elements. In addition to her family and her work, Alonso dedicates time to music. She is a member of a female choir and also enjoys dressmaking. Almost every weekend, she escapes with her family to Extremadura, where the couple has built a hideaway in the mountains, cut off from everything. Here, she relaxes by cooking, spending time with her children, entertaining friends, collecting hazel, and admiring changes in the surrounding nature. She is a natural individual, practical, humane, and straightforward, elements she conveys to each residential project, making her work a benevolent and modern form of architecture.

Portable House ÁPH80

ÁPH80 has three different spaces measuring 291 square feet (27 square meters): a living room/kitchen, a full bathroom and double bedroom. On the interior, its gabled roof is 11.5 feet (3.5 meters) high. Most of the materials used can be recycled and meet the sustainable criteria that ÁBATON applies to all its projects. It blends in with the environment thanks to large openings that bring the outdoors in. The use of wood throughout the building not only adds calm and balance but it is also hypoallergenic. The sourced wood comes from regulated forests (these will regrow to provide a wide range of other benefits such as further carbon storage, oxygen generation, and forest habitat).

The outside is covered with gray cement wood board and a ventilated façade includes 4.7-inch (12-centimeter) thermal insulation around the building. Inside, timber panels are made of Spanish fir tree dyed white. The solid timber structure has been fully designed and manufactured in Spain through numerical control. Manufacturing time takes eight weeks. Assembly time takes one day and transportation is by road. Currently, we are developing a simpler series, which can be added to the ÁPH80 to suit every need, creating larger spaces and contributing to the project's versatility.

The project was awarded 2014 Architizer A+ Awards Architecture + Living Small and Single Family House XS and 2014 Mention Award at the 4th Biennial Iberoamericana of Design.

Photography by Juan Baraja

Estate in Extremadura

Located in a privileged environment in the province of Cáceres, the goal was to transform an abandoned stable into a family home by completely renovating it in a way that would be respectful to and consistent with the surrounding environment.

Problems relating to the electricity and water supplies were solved through the sustainable systems used in all ÁBATON's projects: a solar panel system with storage batteries was developed to supply energy during the summer; turbines were installed taking advantage of the course of the two streams that run through the estate, providing electricity for the winter.

Water elements can be observed throughout the interior of the property, as it flows naturally underneath. Water was also harnessed to become a fountain that is situated in the interior courtyard of the property, while an enormous central lounge takes up the remaining space.

Stone found in the surrounding area was used to build the house and the swimming pool. Materials used throughout the project include recycled stone, concrete, limestone floors, steel, and oak. The interior's supporting walls were replaced by light metal pillars, and the haylofts in the upper area were converted into bedrooms.

This project has been featured in *The Huffington Post* (USA, January, 2012) as an example of a sustainable and off-the-grid house. It is one of the most internationally featured projects in Spain. The project was awarded an Architizer A+ Award for Architecture + Sustainability and the prize "Segnalato in Premio Internazionale," in Italy's *Domus Restauro e Consevazione*, 2013.

Photography by ÁBATON Architectura and Belén Imaz

Las Marias

This house's special design results from the site's unique characteristics and constraints: a desire to respect the existing oak trees, and the lot's distinct layout which dictated that elements of the structure were conceived as a stilt house on a highly resistant granite rock that flies over the terrain and lets it breathe. In order to support the structure, special drilling machines with diamond disks were used to perforate it with extreme precision. This area is home to the sitting room with a stair to gain access to the rooftop swimming pool. A crystal-clear passage links this part of the house with the other.

The house is not just part of its surroundings but dominates them as it can be seen from all angles. Our client was keen to separate the public and private areas. So, the sitting room, the kitchen, and the swimming pool are located on the upper floor and the bedrooms and bathrooms on the ground floor. The owner wanted to give his bedroom and bathroom more value so we integrated both into one single space, using the same materials throughout. A limestone bathtub helped give the space a sense of continuity and the fact that it was built into the floor allows the client to enjoy garden views from every angle.

Materials used in the build include: honed limestone floors laid over an underfloor heating system; plastered walls finished in matt white paint; façades clad with colored mortar render; flat roofs layered with gravel; interior woodwork and doors made from white-lacquered MDF, and exterior joinery made from lacquered aluminum.

The project was awarded The Most Sustainable Housing Development Award 2009.

Photography by ÁBATON Architectura

Vaibhav Dimri
Founder and Partner

K-110
Hauz Khas Enclave
Delhi 110016
India

www.anagramarchitects.com
vaibhav.dimri@anagramarchitects.com
+91 (11) 2652 2137

Our practice endeavors to respond to the economic, sociocultural, and environmental contours of projects. Regardless of scale, architecture has the power to affect lifestyles, respond to aspirations, and reflect social conditions. In order to develop an edifice that resonates with its stakeholders and general cultural values, we interrogate projects parametrically, programmatically, and idiomatically. This helps us sculpt the relationships created by us, between built and unbuilt, inside and outside, humans and nature.

Design and construction are collaborative and nonsequential processes. In most projects, there exists collaborative potential with clients, users, builders, consultants, and technicians. We attempt to explore this to design more holistically within projects' broader objectives. We also undertake self-initiated collaborations with other designers, academics, and creative practitioners to enrich and invigorate our own experiences.

Architecture is more than the creation of structure or real estate. We explore an architecture that is not only physically sustainable but promotes an experiential reconnection with ecology and nurtures responsible and socially conscious lifestyles. Our attempt is not to provide prescriptive solutions but to be descriptive in our processes. We actively engage and communicate with stakeholders about the implications of design decisions during the course of the project.

Anagram

Architects

Gairola House

Our client, based outside Delhi, wished to develop this plot purely for the rental market. Our design avoids the cookie-cutter stack prevalent in Delhi's suburbs and creates an extroverted, multiple residence apartment stack, exploiting the external volumes as shared resources (for light, ventilation, and sociability). The aim was to encourage a socially connected, urbane lifestyle without compromising privacy and individuality.

To maximize rentability, the stack consists of a four-bedroom duplex, a two-bedroom simplex single-floor unit, and a one-bedroom rooftop apartment. These cater to the typical tenant groups of nuclear families, young couples, and single professionals or students.

With these diverse floor plans, the shared courtyard and front setback volumes are no longer simple voids but terraced sociopetal spaces with increased interconnectivity. The open staircase too offers a shared territoriality and enhanced sociability. The sheltered sunken court, cantilevered veranda, and rooftop terrace garden provide each apartment with its own private outdoor space.

The animated play in volumes and voids is further fleshed out with the juxtaposition of materials, cladding details, and textures. The stairwell is designed to be a shared vertical community garden planted with potted herbs, flowering plants, and aromatic creepers leading onto the landscaped roof garden.

Photography by Asim Waqif

South Asia Human Rights Documentation Centre

The South Asia Human Rights Documentation Centre (SAHRDC) is a small, non-governmental organization with limited resources. It required an office to be made on a 538-square-foot (50-square-meter) corner plot emphasizing spatial efficiency and cost-effective construction. Although some fortification against the street was required, it was crucial for the façade to "converse" with external activity. The external wall is conceived as an animated, dynamic skin reflecting the bustle of the street and activating the monolithic façade. The porosity of the wall maintains a degree of privacy while playfully engaging with the street corner.

Spatial efficiency was achieved through consolidating the floor plate to be flexibly partitioned as per the client's requirements. Costs were minimized by using exposed brick construction and by creating a beamless soffit on every floor through a gently vaulting roof. The volume is serviced by a bay of a single-flight cantilevered staircase and a toilet stack.

This buffer bay forms a breathing thermal barrier along the sun-facing side. The porosity of the wall ensures that the buffer bay is well ventilated and yet shaded so as to reduce the amount of heat transmitted to the workspaces. A single repeating brick module laid in staggered courses creates a visually complex pattern and an undulating surface similar to traditional South Asian brises-soleil. The construction of the screen wall was a result of a five-week on-site collaboration between the masons and the architects.

Photography by Asim Waqif

The Digit

Our clients are one of the largest outdoor media companies in India, at a time when billboards make identities ubiquitous to the urbanscape and their clamorous language affects the urban aesthetic. Outdoor advertising relies on marketing through high visual effect in an extremely brief engagement. Our design investigates the notion of identity (through its most common idiom, the thumbprint) and its urban projection. The shimmering, fluttering, red screen perforated with the company's logo is an idiom for its "digital identity", while the identity of the individual is alluded to in the thumbprint ceiling at the reception. We felt that the architecture must project an arresting visual language and the split-second impact of a roadside billboard while simultaneously committing to a deeper value and meaning to the organization, the edifice, and the occupant.

The thumb-like volume reflects reducing floor occupancy with increasing echelons in the organization's hierarchy. It also helps manage daylight and heat gain. The semi-elliptical cylinder creates linear voids, which stretch along the length of the building and infuse the workspaces with greenery and natural light diffused through a skin of louvers. The front façade mitigates noise and thermal gain from the south.

Photography by André J. Fanthome

Yuyang Liu
Principal

No. 9, Lane 281, Wuyuan Road
Shanghai 200031
China

www.alya.cn
office@alya.cn
+86 (21) 54041288

I would describe contemporary Chinese architecture, which has gained momentum thanks to China's economic reform since 1979, as in the state of being overly influenced by globalization. While in search of its own identity at the moment, it will no doubt reshape the future of global architecture and urbanism for the next 30 years.

Personally, I would describe my design philosophy and architectural values as follows: to tackle each project with a sense of sobriety against the desire for feverish growth, to embrace programmatic and urban complexity, and to experiment with an ecological approach in the practice of architecture and urbanism.

I believe for the next stage of young Chinese architects that it is important to return to the fundamental, i.e. observing the intrinsic qualities found in a site or a programmatic requirement, which forms the genesis of a formal and spatial language that can be adapted to a unique architectural strategy. And to understand the way of life of Chinese culture which affects spatial layout, urban layout, organization, and materials, among other things.

Architecture and city both as image and space are much more flexible and mutable for the Chinese. This, I think, is attributed by understanding the world and the environment not as a built reality but as a conceived hyperreality—an imaginary future as reality, if you will. However, the main urban problems facing China today are immense: environmental pollution, social injustice, economic inequality, to name just a few. Fortunately we observe in the Chinese government an increasing awareness on the issues of urbanization, density, and sustainability.

Atelier Liu Yuyang

The avant-garde in the West, has somehow become the mainstream in Asia in the last 10 years. This has a strange effect of popularizing and at the same time trivializing contemporary architecture. The effect on young architects—myself included—working in China as well as globally, is that one loses the traction that is required to produce good architecture. There is no resistance to conservatism or to radicalism, resulting in a kind of zero-gravity context for an architecture that is a free-for-all. Architectural production becomes similar to something like the production of daily news. I like to conclude with a paraphrase from Koolhaas in the *Great Leap Forward* in referring to the Asian urban conditions:

Global presence doesn't equal global knowledge. I think the true value of Chinese avant-garde architects is to go beyond the superficiality of what people see in the Chinese metropolises and to reveal the true cultural conditions of China as represented through its built environment.

No.18 Guan Shu Yuan Hutong

The project calls for the renovation of a small courtyard house in Guan Shu Yuan Hutong, Beijing, which is located between the famed Yonghe Tibetan Temple and the historical Imperial Academy. The client is an art collector from Shanghai, whose wish is to reinvent this "pockethouse" into a gallery with a "Shanghai ambiance" for his ceramic artworks.

My design starts with two parallel aspects: on the one hand, to infill a series of copper plated steel display windows and veneered wood display shelves within the original wooden frames of the house, thus creating an integrated contrast of tradition and modernity; On the other hand, to recreate a courtyard reminiscent of Jiangnan (a region south of the Yangtze) regionalist expression and articulation.

Photography by Jeremy San

Nanjing Road Pedestrian Kiosks, Shanghai

Equipped with hybrid programming, these kiosks take on the notion of "micro-urbanism." Through the carefully controlled formal theme and variations, the 12 kiosks as a whole compose a unique urban landscape. The Art Deco-inspired custom made pattern glass panels echo the historic architecture of the neighborhood. With solar panels on the roof supplying 180 watts per hour of exterior LED lighting, the kiosks become new environmental prototypes.

Photography by Jeremy San and Shen Zhong Hai

Xing Ping Old House (Yun Lu Hotel)

The Yun Lu (Cloud House) is a boutique eco-resort nestled in a village between Guilin and Yangshuo, situated along the dramatic landscape of the Li river. The site consists of five renovated historical farm houses and one new addition which functions as a restaurant and the main gathering space for hotel guests. Taking on a sensitive approach to the local culture with villagers still living nearby, the overall planning and landscape design worked into the original village structure.
I retrofitted the rammed earthed buildings to accommodate refreshing and uncompromisingly contemporary living, while the new restaurant addition adopts an understated presence with the use of steel frame, glass pivot doors and windows, rough-cut stone blocks, charcoal treated wooden louvers, and terra cotta roof tiles.

Photography by Su Sheng Liang

Benjamin Ball
Principal

Gaston Nogues
Principal

2018 N Vine St
Los Angeles
CA 90068
United States

www.ball-nogues.com
inquiries@ball-nogues.com
+1 (213) 4583673

Ball-Nogues Studio is an integrated design and fabrication practice operating in a territory between architecture, art, and industrial design, led by Benjamin Ball and Gaston Nogues. Our work is informed by the exploration of craft.

Essential to each project is the "design" of the production process itself, with the aim of creating environments that enhance sensation, generate spectacle, and invite physical engagement.

Ball-Nogu

es Studio

Rip Curl Canyon

Rice Gallery commissioned this installation in collaboration with The Museum of Fine Arts in Houston exhibition, The Modern West: American Landscape 1890–1950. When the gallery director mentioned a Modern West tie-in before we had settled on an approach to the project, we realized that the notion of landscape and geological phenomena dovetailed with our design for Tiffany and Company's Frank Gehry jewelry launch gala on Rodeo Drive, Los Angeles in 2006. In the Tiffany project, the jewelry maker's "Body as Landscape" ad campaign informed our approach to creating laminated cardboard walls and ottomans. At the Rice Gallery, we expanded the potential of constructing landscapes in cardboard to include the viewer's physical participation. We invited visitor exploration by extending the casual social terrain of the campus into the gallery, transforming it into a traversable rolling playground. On any given day one might discover a group of gallery-goers studying, snoozing, climbing, sliding down the rolling terrain, or even making out in one of the darkened recesses below the cardboard surface.

Rip Curl Canyon was a kind of mythical location in the American West where land and water collide, far from Houston's flat, drained swamps. From its highest point at the rear of the gallery, its steep, crevice-like formations sloped down and gained momentum before breaking apart to form ribbons of curling waves. Like rip currents – narrow, fast moving belts of water – the segments twisted and surged toward the front glass entry wall. The view through the glass provided only glimpses of the unfolding topography beyond and invited the visitor to probe deeper. Steady climbing and exploration by visitors caused the raw cut cardboard to slowly compress with each footstep. Over time this accumulation developed into subtle pathways.

The fabrication processes that we used to make the natural brown surfaces are in the lineage of those that Gehry employed in his legendary *Easy Edges* line of furniture in the 1970s.

Expanding on this knowledge enabled us to create architecturally scaled cardboard structures and introduce double curvature. We used the properties and limitations of the material—determined through building fully scale mockups during development, combined with a parametric digital interface to shape the cardboard ribbons. The project required laminating over 20,000 strips (weighing approximately eight tons) of curved, industrially die-cut corrugated cardboard in 12 days. Incredibly strong, and capable of supporting the weight of several people, the cardboard laminates operate as semi-monocoques with an intermediary plywood armature. The armature was made of standard wood materials – two-by-four and plywood – individually cut and CNC routered offsite to conform to the varying dimensions and curvature of the undulating cardboard shells. We digitally developed a language of slotting connections so that these non-standard parts came together like a giant puzzle in four days, required very little structural decision making in the field, and gave us the freedom to make improvised choices when installing the cardboard.

Photography by Nash Baker

Maximilian's Schell

This vortex-shaped, temporary outdoor installation in the Los Angeles exhibition space of Materials & Applications (M&A), warped the flow of space with a featherweight rendition of a celestial black hole. Hovering over M&A's courtyard, Maximilian's Schell was a spectacle the size of an apartment building constructed from tinted Mylar resembling stained glass. The piece functioned as a shade structure, swirling overhead for the entire summer of 2005. The interior of this immersive experimental installation created a beckoning outdoor room for social interaction and contemplation by changing the space, color, and sound of the M&A courtyard gallery. During the day as the sun passed overhead, the canopy cast colored fractal light patterns onto the ground while a tranquil subsonic drone from the integrated ambient sound installation by composer James Lumb entitled "Resonant Amplified Vortex Emitter" lightly rumbled below the feet of the viewer. When standing in the center or "singularity" of the piece and gazing upward, the visitor could see only infinite sky. In the evening when viewed from the exterior, the vortex glowed warmly while both obscuring and allowing glimpses of the building behind it. The assembly paid homage to a character played by actor Maximilian Schell in Disney Studio's forgotten sci-fi adventure, The Black Hole. Dr. Reinhardt is a visionary tyrant on a monomaniacal quest to harness the "power of the vortex" and possess "the great truth of the unknown."

We invested more than a year in a development process which involved several prototypes, although actual fabrication took only two weeks. The result was an installation that functioned as not only architecture and sculpture but as a "made-to-order" product through a unified manufacturing strategy. We achieved the aesthetic effects by manipulating Mylar reinforced with bundled Nylon and Kevlar fibers on a computer-controlled (CNC) cutting machine. Simultaneously reflective and transparent, the amber-colored film offered UV-resistance through a laminated golden metallic finish. The result was neither a tent-type membrane nor a cable net structure in the manner of Frei Otto, but a unique tensile matrix comprised of 504 different instances of a parametric component or "petal," each cut and labeled using the CNC system. Every petal was connected to its neighbors at three points using clear polycarbonate rivets to form the overall shape of a vortex. As though warped by the gravitational force of a black hole, the petals continually changed scale and proportion as they approached the singularity of the piece.

An integration of structure and skin, the vortex behaved as a "minimal surface": prestressed, always in tension, yet definable mathematically. Its lineage is in the soap film surfaces modeled by Otto in the 1950s and 1960s; a process now typically accomplished using software that performs "finite element" calculations. After we prepared hand sketches and computer models, membrane engineer Dieter Strobel digitally crafted and refined the minimal surface model. He quickly and precisely manipulated it during the "form-finding" process while accounting for the distorting effects of gravity and enabling the finished vortex-shaped canopy to be in tension everywhere across its top surface. This gave it a pure and smooth appearance, especially when viewed from the exterior. Seen from the interior, the piece resembled an enormous transparent flower with its petals lightly draping and curling downward with gravity.

Photography by Scott Mayoral (top left); Neil Cochran (right)

Air Garden

Air Garden embodies the qualities of light and space that are unique to Los Angeles. Like the city itself, it does not have a distinct beginning or end; inside or outside; back or front. It is both an object and an atmosphere. Its appearance is not static as it is predicated on the changing quality of light in the north light well, at any hour of the day. Air Garden is a serene moment amid the hectic action and movement at the airport. Los Angeles airport (LAX) is a city among cities which envelops the globe, networked through a system of concourses, gates, terminals, and connecting flights; it is an international metropolis of movement. Air Garden is a pause within this movement; a place for reflection and repose, an opportunity for the traveler to daydream. Within the confines of passport kiosks, security checkpoints, ticket counters, and other forms of surveillance and control, it is our aim for this work to engender a sense of freedom.

Like most gardens, it cannot be comprehended from a single point of view; and by participating in the ambulatory movement customary to the airport experience; it is inherently linked to the trajectory of one's journey through the City of Cities. Its components are made of gestural volumes of color hovering within an immense array of catenaries. These voluminous brush strokes on a translucent three-dimensional canvas dissolve into washes of color, then snap back into clear strokes with one's changing perspective. They might be perceived as episodes forming an inconclusive riddle without a definitive resolution, even after the complete cycle of one's journey. Our hope is that the Air Garden procures a mental clearing for the voyager which is a respite from the pressures of travel and which produces a felicitous condition for daydreaming.

The work is like a cloud inside the light well, which can be clearly described using a term from chemistry: suspension.

A suspension is a heterogeneous fluid or gas with solid particles more or less evenly dispersed within in it. It is not opaque but more like a ubiquitous fog permeating the space. Therefore, Air Garden does not obstruct the viewers' perspective through the light well nor into the surrounding spaces which showcase the human activities of the airport.

While the environment is interspersed with the metallic bead chain catenaries, it is also constructed from the negative space between the catenaries; sight extends into and throughout the building.

When developing the color composition of Air Garden, we looked at the organization inherent in the architecture. We intend to shape color in three dimensions to echo the architectural order of the Bradley West Terminal by mimicking the shadows cast onto the array of catenaries by the structural members of the light well itself. We chose two dates at which to cast the shadows—the summer solstice and the vernal equinox.

To better understand the effectiveness of the hovering colored forms in the Air Garden, we applied our findings from precursory research on theories of perception. Testing in Gestalt psychology has proven that our mind puts together what is not explicitly present and imposes meaning and structure on visual input. These theories support our assumption that the traveler will mentally connect the architectural order of the building across the void of the light well by way of color painted onto the cloud of bead chain.

For this reason, we are able to achieve a visual connection without having precisely to replicate the geometry of the building but rather to imply its presence.

The reflective qualities of the bead chain, which forms the catenaries, create a sense of vastness through the play of light in space. Each one is a miniature convex mirror, capturing light from all corners of the space as well as from the adjacent balls. The project evokes a sense of immensity through the reflective dance of color and light. The consistent repetition of highlights on each bead produces a condition of being ensconced in a place that is both proximate and seemingly vast.

Los Angeles is a region full of creative opportunity and innovation; Air Garden represents these characteristics.

Here, creative economies thrive. The real meets the unreal and Air Garden is a metaphor for this through its presence as both painted gesture and atmosphere. Much like the infinite space in a traveler's momentary lucid daydream, Los Angeles is a boundless expanse. Our city seems limitless when coasting its highways' smooth curves, which pattern its extensive landscape beautifully with measured grace and elegance. Air Garden embodies the sublime qualities of the utopian paradise we call home, Los Angeles.

Photography by Joshua White

Herwig Baumgartner
Partner

Scott Uriu
Partner

834 S. Broadway
Suite # 502
Los Angeles
CA 90014
United States

www.bplusu.com
contact@bplusu.com
+1 (213) 6232347

Herwig Baumgartner and Scott Uriu of Baumgartner+Uriu (B+U) are an internationally recognized design duo operating at the forefront of contemporary design. Our design process can be described as driven by digital techniques and advanced computation that utilizes new technologies and material resources. Baumgartner+Uriu's work consistently pushes the boundaries of architecture and urban design by experimenting with new spatial concepts, intensifying existing urban landscapes in pursuit of a visionary aesthetic that encompasses all fields of design. Our architects' work has comprised cultural locations including museums, concert halls, and exhibition spaces; educational and transportation facilities, master planning and urban design, offices and mixed-use developments, restaurants, and residential work.

Baumgartner+Uriu's work was recently exhibited at the FRAC center in Orléans, France; the Museum of Contemporary Art (MOCA) in Los Angeles, California; the Centre Pompidou in Paris, France; the 12th Architecture Biennale in Venice, Italy; the Architecture and Design Museum (A+D) in Los Angeles; the SCI-Arc gallery in Los Angeles, California; SITE Gallery and David Richard Gallery in Santa Fe, New Mexico; INSITE Gallery, Bangalore, India; the *2011 AIA Emerging Professional* exhibition in Washington D.C.; the *Stadtkrone 2030 Exhibition* in Milan, Italy; the University of Applied Arts in Vienna, Austria; the Southern California Institute of Architecture (SCI-Arc); and the Los Angeles Forum for Architecture and Urban Design in Hollywood, California.

Baumgar

Two monographs have been published on Baumgartner+Uriu's work. The most recent one, *B+U- Designpeak 12*, was published by Equalbooks (Korea, 2012) as part of its monograph series. Baumgartner+Uriu's work has also been widely published and discussed in books magazines and newspapers such as *Architectural Record*, *The Los Angeles Times*, *Los Angeles Magazine*, *Azure*, *FORM*, *Mark* magazine; as well as on television and radio.

Baumgartner+Uriu is an award-winning firm, having received the Maxine Frankel Award for design research (2010), the AIA National Award for Emerging Professionals (2011), the A+Award for sustainability (2012), the Graham Grant for Advanced Studies in the Fine Arts (2014) and the COLA Fellowship for 2015 from the City of Los Angeles – Department of Cultural Affairs.

Herwig Baumgartner, an Austrian native, is a licensed architect and principal/co-founder of Los Angeles-based Architecture firm Baumgartner+Uriu (B+U). He received his diploma in Music and New Media from the University of Music and Performing Arts in Vienna and his Master's degree in Architecture from the University of Applied Arts in Vienna. Baumgartner has lectured and conducted workshops nationally and internationally and is a professor teaching design and applied studies in the graduate and undergraduate programs at the Southern California Institute of Architecture (SCI-Arc). Most recently, he represented SCI-Arc at one of the leading symposiums on the latest developments in architectural geometry and computation at the Centre Pompidou in Paris. Prior to founding B+U, Baumgartner worked as a senior associate and project architect at Gehry Partners and also collaborated with artist Richard Serra.

Scott Uriu is a licensed architect and principal of B+U, which he co-founded. He was born in Davis, California, and was initially an undergraduate mathematics major at the University of California, Davis. He then received his Bachelor of Architecture degree from California State Polytechnic Pomona and studied at the Architectural Association in London, Diploma Unit 7. Uriu has lectured and conducted workshops nationally and internationally and is a professor who teaches design studio and applied studies at the University of Southern California (USC) and the Southern California Institute of Architecture (SCI-Arc). Previously, he was part of the design faculty at Woodbury University. Prior to founding B+U, Uriu worked as an architect at Gehry Partners.

tner+Uriu

Frank and Kim Residence

The Frank and Kim Residence is located 9 miles (14.5 kilometers) north of downtown Los Angeles on a hillside property in Pasadena. The existing two-story single-family house is placed at the end of an elongated property, on the edge of the hillside with a large terrace and a pool overlooking San Marino. The main feature element for the front yard is a large canopy to be used for receptions and social events, which articulates a new dynamic entrance to the house. The 65-foot-long (20-meter-long) cantilevering steel structure is clad with a white translucent fabric. The canopy is lit from the inside with a combination of white and colored LED light fixtures to illuminate the garden with a soft glow during the evening, articulating a vibrant path to the building and marking a spatial continuity throughout the site. As part of the design process, we developed many iterative models using a continuous folded surface geometry that emerges from a single plane and through techniques of splitting (delaminating) and shifting creates a multilayered volume with varying translucent qualities. We were interested in the effect this geometry had on the gradation of light. Without changing the material property we were able to produce the desired light quality from the inside utilizing white and colored LEDs as a light source.

Photography by Josh White

Taipei Performing Arts Center

The design proposal for the Performing Arts Center in Taipei creates a world-class institution, which is characterized by both its response to its urban and cultural environment, and by its formal and structural elegance. The project embraces the concept of a Grand Plaza as being a central hub between the Shilin Night Market and the TRTS Jiantan Station. This is achieved by lifting the multiform theater off the ground and creating a covered outdoor linkage space between Cheng De Road and Wen Lin Road. This linkage space is the central access to the grand foyer and all three theaters.

On the south side, the Grand Plaza transitions into a garden that gently slopes up towards Jiantan Road to allow for shops and restaurants to be placed underneath, making it accessible from street level. Both the garden and the Grand Plaza provide a large outdoor space for the thousands of pedestrians who visit this area. It is lined with shops and restaurants at ground level and covered by the theater's large roof above.

The shops and restaurants are located along Wen Lin Road and Jian Tan Road and, together with the Grand Plaza and the garden space, embrace Taipei's lively streetscape by creating a continuous experience from the busy streets of the Shilin Night Market to the Performing Arts Center and the train station.

The Taipei Performing Arts Center is intended to become an important landmark in the region, which expresses the richness and diversity of performing arts and creates a destination point for the area. For the design development, we used a unique process utilizing actual sounds as its basis. The morphology and shape of the building was designed using sound waves that were analyzed and transformed into three-dimensional vectors. These vectors became the formal and structural framework for the design of the exterior envelope. The building materializes with a metal and glass enclosure that reveals its activities in a variety of scales and angles to the city.

Photography by Baumgartner+Uriu

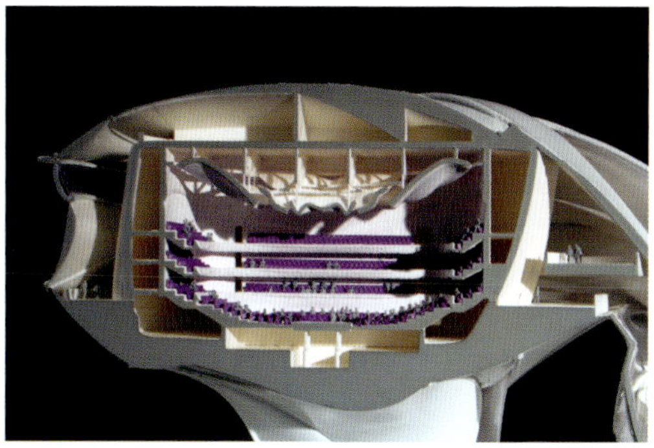

Apertures

Within the discipline of architecture, the discussion of fields, networks, and smooth transitions has dominated the dialog over the past 15 years. Rooted in philosophical models as articulated by Deleuze, systems theory, and parametricism, it has influenced many generations of architects.

Parametricism promotes a relational ontology in which entities have no autonomous reality and are based on "continuous differentiation;" everything is connected, everything flows.

This position of an architecture rooted in dynamism and deterritorialization is being opposed by a radically different approach, giving way to a contemporary design practice working with discrete figures that cannot be entirely understood through its pristine digital relations. This position is one that is obsessed with capturing qualities that would appear to be incongruous, by incorporating analog features into a digital design process. The installation *Apertures*, designed for the SCI-Arc Gallery, is firmly positioned within this approach.

Apertures are the architectural catalysts for the installation design, being defined as objects within a larger building object that differ from its host in terms of morphology and performance. They are disruptive features to the overall building mass, but are able to interact with their environment, focusing on a symbiotic relationship between nature, building morphologies, and material expression. Apertures have been an ongoing topic in our work, challenging the notion of an architectural opening as a static object by re-defining the DNA of a window, both in terms of its appearance and materiality, as well as its nature as an object in continuous flux, responding to its environment through movement or sound.

The 16-foot-high (5-meter-high), thin shell structure was designed to rely solely on its extremely thin surface (0.1 inches or 3 millimeters) as support, requiring no additional structural elements. Structure and surface are collapsed into a single component supported solely through its shape, creased surfaces, and material strength. Each one of the 233 panels is unique in terms of its shape. They are CNC milled from polyurethane foam, heat formed out of thermoplastic polymer resin, and then laminated together into a single object.

Unique to this project is the proposal of building as an organism, challenging how architecture can interface with its users and its environment in a much more intuitive way. This entails both the use of technology to augment its performance and a design aesthetic that is incongruous and can incorporate analog features into a digital design process. The project emphasizes an architecture between nature and technology, which can operate as an interactive building organism where multiple discrete features work simultaneously and independently. In this case, sound is used to bridge the gap between the natural and the artificial, allowing the visitor to experience their own biorhythms.

Photography by Josh White

Bruno Campos
Principal

Marcelo Fontes
Principal

Silvio Todeschi
Principal

Rua Raul Pompéia 225, São Pedro
Belo Horizonte
Minas Gerais 30330-080
Brazil

www.bcmfarquitetos.com
info@bcmfarquitetos.com
+55 (31) 3281-2707

Established in 2001 by partners Bruno Campos, Marcelo Fontes and Silvio Todeschi, BCMF Arquitetos has been involved in the development of a broad range of outstanding projects, and is acknowledged in Brazil and abroad as one of the leading practices of a new generation of Brazilian architects. Since 2010, BCMF shares workspace with two other studios, friends and frequent partners MACh Arquitetos and Hardy Design. The group strives to work collaboratively on urban and architectural projects of various scales and complexities, functioning as a hub of independent companies with different expertise, sharing knowledge and resources. The studio seeks to be highly diversified in terms of scale and typology, ranging from interior design to large-scale projects including multi-purpose arenas, sports facilities, mixed-use projects, hospitals, industries, parks, and urban master plans.

Among the practice's main projects, the multipurpose Deodoro Sports Complex (National Shooting, Equestrian, Field Hockey, and Modern Pentathlon venue) which was built for the Rio de Janeiro 2007 Pan American Games, has received high international recognition. This includes a Gold Medal at the 2010 IAKS Latin and Caribbean Awards and a Special Prize at the IOC/IAKS 2011 Award in Cologne, Germany. Subsequently, in 2009, BCMF Arquitetos was selected by the Brazilian Organizing Committee to develop most of the concept studies for the winning Rio 2016 Olympic Games Candidature, during which Rio competed against Chicago, Madrid and Tokyo. The Dossier of the Brazilian candidature was marked by the IOC (International Olympic Committee) as the one of the "highest quality" among the four competing cities.

BCMF A

In 2010 the studio was commissioned by the Minas Arena Consortium to carry out the renovation of the legendary Mineirão Stadium in Belo Horizonte, one of the selected venues for the 2014 FIFA World Cup and the first stadium in Brazil (second in the world) to receive LEED Platinum certification. As a result, BCMF became the first Brazilian architectural practice to participate in the planning of the three major mega-events that have promised to transform the country: the Pan American Games (2007), the FIFA World Cup (2014) and the Olympics (2016).

In 2014, BCMF Arquitetos was listed by Fast Company as the second most innovative company in South America.

Bruno Campos is an architect graduated from the Universidade Federal de Minas Gerais (UFMG) in Belo Horizonte, Brazil in 1994, followed by an MA in Housing and Urbanism from the Architectural Association School in London in 1998. He collaborated with various architects in Brazil and New York before establishing BCMF Arquitetos in 2001, together with Marcelo Fontes and Silvio Todeschi, also graduates of UFMG. Since then, they have designed a vast array of projects, varying both in typology and scale, but especially large-scale public projects. The practice's work has been widely exhibited and published in Brazil and abroad.

quitetos

New Mineirão Stadium

Inaugurated in 1965 (original design by architects Eduardo Mendes Guimarães and Gaspar Garreto) as the second largest stadium in the world, the Mineirão Stadium is located by the Pampulha Lagoon, an area which hosts many of Belo Horizonte's cultural landmarks, and close to Oscar Niemeyer's cultural landm's seminal work. In this context, to transform the legendary stadium into a modern, multifunctional facility, the interventions proposed were both radical and respectful, reinforcing the monumental original structure within its iconic modernist landscape.

As for the original architecture of the stadium, only the outer "shell" remained: the 88 structural semi-porticos, the concrete roof, and the upper tiers. We rebuilt the rest of the "core" to guarantee the total overhaul inside the arena, including the new extension of the roof, all the new program and infrastructure, besides the lowering of the pitch and the redesign of the lower tiers, improving the sightlines for the new capacity of 62,160 seats.

Outside the stadium was completely renovated too, with a new 2,153,000-square-foot (200,000-square-meter) operational platform separating the spectators from accredited flows. The "Esplanade" includes various facilities around the stadium, which has been opened to the public as an immense landscaped plaza, visually linked to the Pampulha Lagoon. This platform is molded to the site topography and integrated with the immediate surroundings, being perceived as a continuation of the street domain.

Photography by João Marcos Rosa/Portal da Copa (top); Leonardo Finotti (bottom left); Joana França (bottom right)

Deodoro Sports Complex

The Deodoro Sports Complex was designed for the Rio 2007 Pan-American Games, including in the same cluster the Shooting, Equestrian, Archery, Hockey and Modern Pentathlon facilities. All venues meet international standards, and will need just minor adjustments and complements for the Rio 2016 Olympics. The cluster is already a world-class legacy, which has successfully triggered the renewal and further development of this important suburban region.

The project deals with the complex issues of a unique suburban context comprising a military neighborhood, a densely populated favela, an industrial area and a vast wild landscape all mixed together. The Shooting Center is probably the cluster's most impressive venue, with approximately 538,200 square feet (50,000 square meters) of built area landscaped on a 1,615,000-square-foot (50,000-square-meter) trapezoidal site, along one of the most important access vectors of the city (a hybrid of avenue and motorway). Few materials, strong horizontal lines, and an enigmatic grid of wooden baffles predominate in this stark venue, located on a breathtaking valley surrounded by mountain peaks.

Photography by Bruno Carvalho (left); Leonardo Finotti (below top, below middle and below bottom)

Casa Cor Bar 2010

The brief required us to explore the formal possibilities of applying a very simple and inexpensive construction system: the light steel frame. Industrialized, resistant, high-quality materials (stainless steel plate, mirror finishing on the outside and matte on the inside) clad this custom-designed object built in just three weeks. With dim lighting and some cork benches distributed at strategic points, the orange carpet floor unfolds into bleachers and stepped platforms where people can relax and admire the landscape from inside or outside.

This unusual and irregular cruciform structure looks like a bolide, which has landed on a downtown building rooftop. It was designed to be portable and usable in various ways and can be removed later on to anywhere else without losing its visual impact: to a park, a plaza, a garden, or even to the roof of another building.

Photography by Leonardo Finotti (right top and bottom); Jomar Bragança (far right)

STELLA ARTOIS

Prof. Merritt Bucholz
Principal

Mountpleasant Business Centre
Mountpleasant Ave
Dublin 6
Ireland

www.bmcea.com
mbucholz@bmcea.com
+353 872605851

In the practice I share with Karen McEvoy, we started upon winning an international open competition for the design of a civic office building, Fingal County Hall, in Dublin, Ireland. From our office in Dublin we have developed a civic approach to architecture that is technologically sophisticated, low energy, and oriented to human use and understanding.

Our work has pursued the idea that within architecture, design, technology, and landscape are situated the parallel and evolving structures of human culture and use, and natural ecology. Buildings are the active ingredient among the spheres of climate, landscape, and use. We believe that architecture is the crafting of connections between these parallel structures, and the tuning of the man-made world into climate, environment, use, and construction. We take a human approach to designing low energy architecture in order that buildings are legible and evaluated in human terms as environments rather than the narrow technological and empirical metrics that seem to govern so-called "green" architecture. The refinement of architecture is really about clarifying the performance of architecture, making obvious and human how things operate and to what spaces and materials humans orient themselves. Often, invisible physical things such as gravity and air movement become part of a critical architectural form at the scale of a landscape.

To achieve our architecture we believe that we must work in intensely interdisciplinary teams which we lead on all of our projects. We have had the good fortune to collaborate with some of the world's best engineers such as RFR, Transsolar, ARUP, and Buro Happold. This collaborative practice explores how architecture is

Bucholz McEvoy A

the cultural consequence of the conversation between architecture, structural engineering, and climate engineering. The transparency of the design process we conduct allows those conversations to manifest themselves in the building within an aesthetic framework, so that elements are shaped by this discussion as it traverses spatial, material, empirical, and regulatory concerns.

Teaching is an active part of my practice; to that end I founded a new school of architecture at the University of Limerick (SAUL) in 2005. The school is based on the concept of a "soft-matrix" – a flexible logistical structure that allows students and staff freedom for thought about architecture. This occurs within a carefully constructed design studio, which I designed as part of the building of the school, propelled by the collective work of students and staff working directly with the landscape of issues and territory that surrounds them in what is a regional location.

Staff, traveling practitioners of architecture, writing, engineering, and philosophy, travelling to SAUL from around the world—provide the dense pedagogical framework that student work is the primary catalyst for. This pedagogical reciprocity between students and teaching practitioners produces a constantly new energy around architecture.

Limerick County Hall

View from southwest looking northeast (right); West façade (far right); Civic atrium (far bottom right)

Photography by Michael Moran

Elm Park

Hotel, hospital offices and landscape looking south (above); Office building, canopy and landscape (far left); Apartment building and landscape (left)

Photography by Michael Moran

Fingal County Hall

Civic entrance (top); Council chamber (above); Civic space (left)

Photography by Michael Moran

Carlos Arnaiz
Principal

55 Washington St
Suite 701
Brooklyn
NY 11201
United States

www.cazarch.com
carnaiz@cazarch.com
+1 (212) 7962770

We have started to doubt the innovation narrative. What if our thirst for the cutting edge is misguided? What if all this running does not bring us to the front? What if the nature of the new is not about disruptive innovation but about connections with people and places other than ourselves? What if the importance of design is in its ability to foster exchange instead of revolution?

Our questions are driven in a large part by our need to survive. We reside in a converted warehouse in Brooklyn and work on projects in Manila, Hangzhou, San Jose, Cebu, Lima, and Bogota. The geographical gulf in our work cultivates a necessary fondness for resilient identities. We do not comprehend most of the languages we hear so we must listen closely, make things up as we go, and relax.

Our projects are conceived as an ongoing archive of the foreign bodies we encounter in everyday practice. We invite more associations in our work than we can realistically handle because overlapping ways of thinking are more interesting than an ideology that must be continually transgressed. We attempt to build moments of conviction upon a fragmentary universe of fickle consumer trends, friction-free markets, political unrest, and mind-numbing technological advances.

CAZA

We are drawn to the aberrant situations and alternate rationalities in a project. We spot an imbalance in our environment and produce a story about it: a setback dictate that exacerbates density, a curtain wall constraint that makes us rethink thinness. Our designs seek to refract these existing conditions and nest commentary in architectural form. Our formal accretions become a tacit knowledge of these places and their communities.

We are inspired by the many imperfect visions of survival on our endangered planet. The reality of environmental catastrophe is indeed frightening and if innovation is a form of exchange then perhaps we can design spaces that we have not yet seen – spaces in which irreconcilable desires co-exist. Architectural form could then be a tacit knowledge of our responsibility to dream as a community.

This is not a light task. It comes with the duty of thinking about design as a way of knowing-in-group. As we sit together in the darkness holding hands with strangers, geometry is our campfire hymn. We use measurement to manipulate, influence, and affect form. Rationality is sensible only as a system that we have incrementally translated.

Perhaps this is a foolhardy goal for a small studio in the age of corporate behemoths.

At CAZA, we all share a tenderness for these fledging realities in the form of that "very odd family of deer," as Dr Seuss wrote, "whose horns are connected, from one to the other, whose horns are so mixed they can't tell them apart, can't tell where they end and can't tell where they start!"

We have posed ourselves the challenge of making environments out of a deep affection for that which we don't control. Our romance with the idea of a manifold and optimistic future breeds an epistemological seriousness with regards to our methods. We must train, like amateur boxers, in the absence of our opponent and yet remain committed to their presence. We are natural adversaries who will fight for the chance to recognize the other person in a ring and share a delightful exchange of blows. The preservation of this wild meeting with the unknown is, as designers, our gift to the world.

Costa Rica Congress Hall

Ranked the most peaceful country in the continent, Costa Rica simultaneously supports a loud opposition. The design aims to capture this democratic paradox. The building is organized around a series of structurally dependent and interconnected hypercubes. A platform for public demonstrations arises from the tension between the stability and precariousness of the structure, while its identity emerges out of the simultaneous monumentality and ephemerality of the image.

La Salle Church

The design of the church plays with the idea of boundaries, between the lightness of the porous skin and the weight of the sanctuary, to restructure the relationship between the church and its environment. As part of a larger university master plan, the church is an integral component in weaving culture and ecology through the site. An array of tangent circles creates indoor and outdoor pockets for liturgical functions that vary in spatial constraints and degrees of formality.

100 Walls Church

Like a dense mass of buttresses, ribs, vaults, and spires, the 100 Walls Church retools the early gothic church in the context of the contemporary grid. In spite of a glut of typological clues we chose an "anti-form." We did not want legibility. We sought to reinforce the experience of the search. The multitude of doors and passages is a reminder that there are as many paths as there are lives and that a sacred space today should draw out meaning in its inscrutability.

Photography Iwan Baan

Ali Rahim
Director

Hina Jamelle
Director

233 Spring Street
Suite 801
New York
NY 10013
United States

2200 Wenxiang Rd
Songjiang District
Shanghai 201620
China

www.c-a-p.net
info@c-a-p.net
+1 (212) 366-9333 (New York)
+86 (21) 3701 6333-1006 (Shanghai)

Contemporary Architecture Practice has established an award winning profile in futuristic work, using cutting edge digital design and production techniques. Since the inception of our office, we have been at the forefront of designing projects that leverage the most advanced technologies available to architects. We are constantly active in inventing design techniques, experimenting with the latest fabrication technologies, and incorporating innovative business models in designs. Our goal is to continue to stay at the forefront of innovation in architectural practice. We take pride in the results of this diligence: consistently intelligent, beautiful designs. *Architectural Record* awarded us Design Vanguard 2004 as one of 11 practices building the architecture of the future.

We have recently been published in: *Room* (Phaidon Press, 2014) which features 100 of the most creative interior design projects selected by 10 acclaimed curators from around the world; *Architecture Record's Interiors* (2010); *10x10_2* which presents 100 of the world's most exceptional emerging architects, selected by 10 internationally prominent critics, architects and curators (Phaidon Press, 2005).

The work of the firm has been described by Zaha Hadid as "Elegant ... and [designed by] virtuosos who are using the computer for the most beautiful compositions ... The designs are immaculate, crisp and coherently detailed."

Contempo
Architectu

Beyond the austerities of digital techniques, elegance concerns refinement, precision, and formal opulence. It integrates an aesthetic desire, unleashing a visual intelligence pertinent to all design fields at all scales. The concept of elegance has the ability to push forward the discourse of contemporary architecture by accepting that complex architectural compositions require an accompanying visual aesthetic as sophisticated as the current techniques used to generate form.

Elegance mediates and enables complexity. A tightly controlled, precise refinement in technique is required to mold transformative surfaces that incorporate distinctly different topological features. The results are potentially chaotic. Negotiating and restraining the visual opulence of these compositions is an operation that requires elegance.

An increasingly prevalent trend is the development of digital models which, in addition to incorporating formal constraints, integrate aspects of material and cost in one seamless model. Here, design intelligence moves beyond effective use of the CNC mill or laser cutter towards a highly integrated model which can compute costs dependent on factors such as material curvature and joinery. The customization of these parts and their modulation, fabrication and assembly provide a bespoke quality to the architecture. Thus, the formal attributes of architecture developed during the design process may integrate nuances of negotiation with fabrication industries, directly reducing the cost of mass customization.

We create designs through digital models which employ complex information and incorporate real-time feedback. We guide the design as the number of client parameters increase, incorporating new data such as project goals, site restraints, circulation, materials, and light into an evolving, iterative design. By using these techniques in tight collaboration with our clients and consultants from the onset of the project, we develop unprecedented, singular built forms which seamlessly and sustainably incorporate form, structure, and building systems.

We think that it is imperative to move beyond digital technique and towards an aesthetic and experiential sophistication. We take an approach not concerned with technology for technology's sake, but which sees technology as a tool to help tell captivating stories in which the supporting technology goes completely unnoticed. At Contemporary Architecture Practice, technology is the beginning of a design process which develops the far reaching innovative effects a project can have on its users.

We believe that our innovative design approach, advanced technical expertise, and experience as leaders in architectural design make us the ideal candidate to develop new visions for projects of any scale, from interiors like IWI Orthodontics to masterplans such as Samsung's Raemian Housing.

rary

re Practice

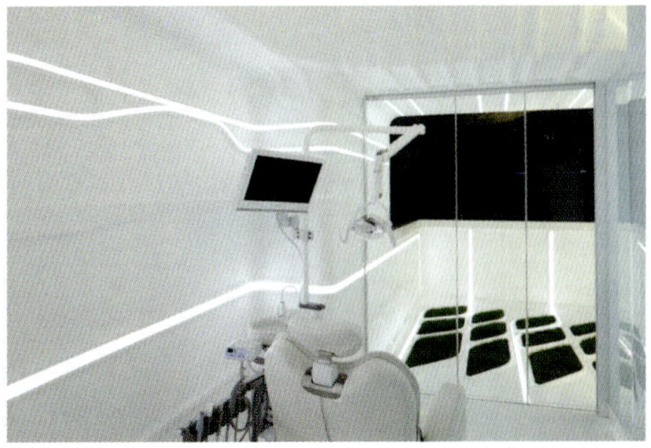

IWI Orthodontics

The first phase of this project is a VIP orthodontic clinic in the Omotesando district of Tokyo, a vibrant and sophisticated area visited by both locals and tourists. IWI uses a new invention to straighten teeth invisibly within 14 months, with one incision in the palette. The clinic is structured using steel and panelized with calcium silicate panels which are free to move in the vertical dimension in case of earthquakes, and which simultaneously operate as cavities for LED lights.

Space is very limited in Tokyo, so we used a strategy to increase the clinic's scale by incorporating the exterior balcony space such that it participates in the interior experience. We develop smooth and gradual transitions between materials—soft to hard, including paving, grass, soft paneling to fabrics, opaque to clear glass, controlled LED/halogen to natural light, while fulfilling the experiential goals of the project—at first startling and then providing a more intimate personalized experience of the space.

The clinic is constructed using panels that are different in design yet they all needed to align perfectly because of the continuous seaming running through them. To achieve this we used real time feedback in the construction process. The panels were fabricated and installed, and if adjustments were required in the field, they were communicated to New York where we altered the parametric model and sent the fabricator in Japan updated files, which were milled and installed. The precision and refinement in this project are possible only thanks to CAP being in control of the entire process from design to manufacture.

Photography by Nacasa and Partners

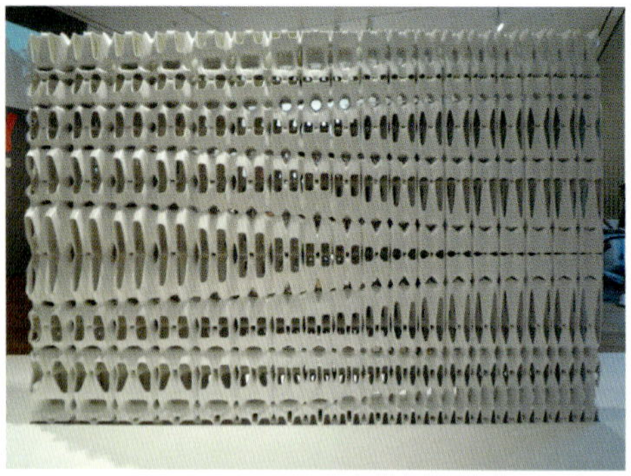

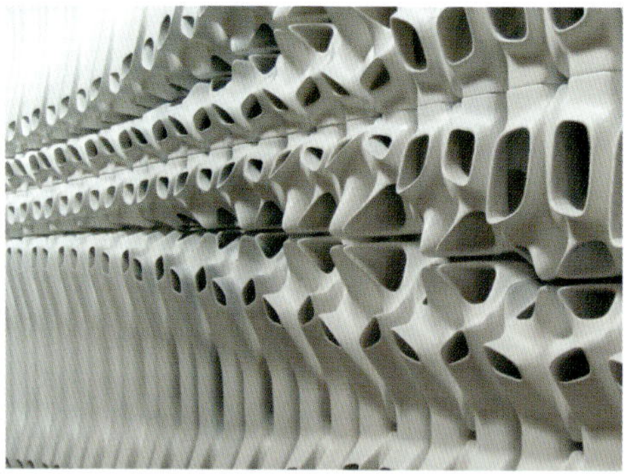

Museum of Modern Art [MoMA] Wall of the Future: Migrating Formations

An installation commissioned by the Museum of Modern Art, New York for their exhibition titled *Home Delivery: Fabricating the Modern Dwelling*, the project was to design a "wall of the future" and the dimensions of the project were given to us by the curators. The wall was constructed by robots using epoxy polymer materials which could structure themselves with all the systems integrated into the form. We think this is the future of residential construction.

Migrating Formations was constructed in two days without any framework, molds, or fasteners, and installed in an hour. Space, structure, and skin are integrated in a single form.

Each side produces different qualities—from bulbous to angular features. The integration of these qualitative differences is controlled by our desire to develop the greatest sensorial effect while yielding an elegant aesthetic.

Photography by Contemporary Architecture Practice

Samsung Raemian Master Plan

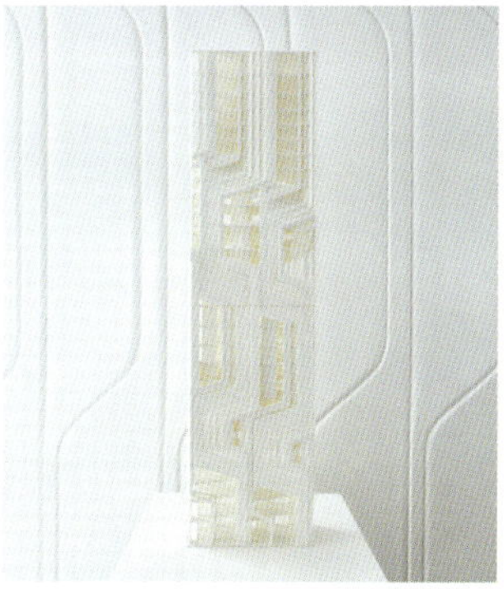

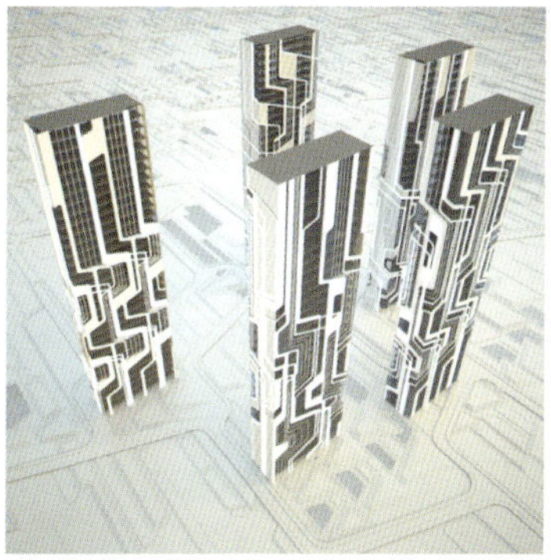

This project is a 36 story master plan and building design for Samsung in Seoul, South Korea, located on the Han River. The master plan incorporates all the programmatic and zoning codes required to maximize views and quality of light incorporating the mid-high rise development. The proposal re-assesses the value of a central recreation area for the entire community and develops several micro-communities around smaller zones for recreation. Benefits are gained, including the ability to provide variation within a micro community through different qualities of buildings and landscapes, as well as the overall master plan. The buildings adhere to strict building regulations while maximizing the variation of views and interior space within the plan configurations. The building panels and glass are flat yet they articulate movement across the façades, enriching the project form.

Samsung needed help in developing a cost effective method to develop variation in buildings in the Raemian housing master plan. We developed an alternative that adheres to Samsung's strict rules of planning configurations, with variations in the floor plans and building elevations. We provided variation by leveraging all of Samsung's research on housing with strict zoning regulations in Seoul, and combined it with our knowledge of writing a design program in C++. We incorporated previous knowledge that we developed with Wharton Business School, which generates costs as each master plan is generated in real time.

Photography by Nacasa and Partners

Hitoshi Saruta
Architect

661-123 Higashi Koiso
Oiso-machi
Kanagawa-ken 255-0004
Japan

www.cubod.com
cubo@cubod.com
+81 (463) 74-6932

We aim to create structures and spaces with a strong contrast of light and shadow, which add beauty and sustainability for years to come. We focus not only on versatility, but on how to appeal to one's visual sense and evoke reflection. I've sought, in my life, to look deeply within in order to discover what is truly most interesting to me. This is more difficult than it sounds, as it requires direct engagement with one's fears, insecurities, and assumptions. Then, whatever it is that I've found, I seek it out in the world and put myself in this situation.

CUBO design a

rchi ect

2KAM

This residence is located in the mountains, far removed from the hustle and bustle of the city.

To truly experience the fine view, the windows in the living room on the second floor can be fully opened.

Though the layout may appear to be simple, residents can appreciate the texture of the stones and wood, and are provided with an atmosphere of luxury and sophistication in harmony with nature.

One can use all five senses to appreciate the nature and tranquility of Kamakura, especially the sense of smell as wood burns in the fireplace.

Photography by Yasuno Sakata

cnest

This is a plan for a site, which can be called a cliff with a maximum elevation difference of 14m and a maximum slope angle of about 70 degrees. The biggest draw of this site is the views of the sea horizon and rich greenery. The house was planned with consideration of maximizing the utilization of its appeal. An intuitively germinated image there was of a floating birdhouse with a pointy delta-shaped roof, hanging from a big tree. The cantilever with a 14 x 43-foot (4 x 13-meter) flat surface, which supports the floating birdhouse, is supported by an RC slab and narrow steel-frame branches.

The way of letting natural light into the interior space allows one to feel the purity of the sunlight, as well as intensifies the beauty of the shadows, which is at the extreme opposite of the light. Walls are arranged to dramatically invite the strong light. A gradation of shading, which is visualized by trimming and manipulating light, causes retinal shrinkage as well as reminds us of the traditional Japanese appreciation of "shadow" parts. Depending on the amount of light, grace and modesty are born in the space and the space becomes dignified and silky.

Photography by Hiroshi Ueda

SOL

While drafting the design of cnest, we received an offer to design a house on the neighboring lot by sheer coincidence.

Although the layouts are quite similar to each other due to similar terms and conditions of construction, SOL has defining characteristics of embracing light and framing scenery.

Designed using sequential repetition, the unusual arrangement of large square windows provides residents with a breathtaking view from almost every angle. Whether sitting, standing, or walking, one will have a picturesque view at any given hour of the day.

We have designed a quiet house filled with light by once again controlling the contrast of light and shadows.

The views are equally beautiful at dusk, as one has an impressive view of the sun melting into the lush landscape surrounded by the Oiso Mountains.

Photography by Yasuno Sakata

Damith Premathilake
Principal

40/A Veediya Bandara Mawatha
Ethul Kotte
Battaramulla
Western Province
Sri Lanka

www.damithpremathilake.com
damithcp@gmail.com
+94 77 1960980

Damith Premathilake Architects is a design firm established by award-winning architect Damith Premathilake in 2008. Since its formation, my firm has evolved into an inventive and energetic team of young architects, consultant engineers, landscape architects, and furniture designers. We were the winners of the Design Award of Excellence at the annual awards ceremony, conducted by the Sri Lanka Institute of Architects in 2012.

We have continued to grow into a formidable design firm. Winning the World Architecture Community Award in 2012 and 2014, as well as the Architecture Asia Award for Emerging Architects in 2014 elevated us to one of the most highly decorated design firms in the country. My firm specializes in personalized houses, office buildings, commercial buildings, restaurants, hotels, hospitals, urban design, and town and country planning, as well as landscape and interior designing. Through this varied cross-section of projects, we try to find a symbiosis between aesthetics, economics, functionality, and climate.

Damith Prematihilak

I strive for perfection in terms of functional creativity, providing unique solutions while aiming for client satisfaction. My designs represent volume in relation to context, aesthetics, and vivacious internal arrangements, which induce positive user behavior. Designed with careful consideration to artificial elements and natural landscape, my designs are personalized to achieve optimum benefits, despite the cost and site limitations of individual projects.

Together with our team of consultants, my firm offers consortium services required to successfully complete building projects. Apart from architectural, interior design, urban designing, and landscaping services, we offer services from associated disciplines such as structural engineering, electrical engineering, water supply and drainage, and quantity surveying and project management, undertaken by well-qualified and experienced professionals who are experts in their field.

e Architects

Holiday Home at Maduru Oya

I led the renovation project at Maduru Oya Army Camp, where the training school for Special Forces of the Sri Lankan army is located. The camp is surrounded by jungle and faces the lakefront of Maduru Oya.

Maduru Oya lakefront was the backdrop to the site with a mountain range establishing the contextual enclosure, stretching toward the land side of the camp. The visual axis toward the mountain range takes the visitor's mind into different settings with the water views of the lakefront and the mountains forming a green belt under a canopy of blue sky. It is an inviting place to relax. While the client and I were having a discussion about the lake-front renovation project, the client remarked on the tranquility of the setting.

At the site, we also found some container boxes abandoned after the war. These container boxes with their enormous interior space provided me with an excellent opportunity to design a setting using them as part of the created space.

The design captures a panoramic view of the rugged wilderness as well as the scenic setting of Maduru Oya Lake and the blue mountains behind. Not forgetting the picturesque sunrise over the mountains.

Photography by Logan McDougal Pope

Filling Station at Ampara

When undertaking this project, we encountered several challenges but created an elegant structure, which will definitely be a little landmark in this suburban landscape. A fact already proven as its 3D design has been selected as one of the "Top Five Modern Filling Stations" in the world.

The narrow and flat site is located along the Ampara to Kandy highway 0.6 miles (1 kilometer) away from Ampara, a remote city situated on the east coast of Sri Lanka.

The design is sculptural in its nature and yet thoughtful in its detailing. Despite being constructed predominantly from concrete, the structure appears delicate due to the strategically placed voids in its façade.

The scheme combines commercial and residential functions and places them under one architectural umbrella. The beauty of the project lies mainly in the simplicity of the planning. The ground level functions as a service station while the floors above facilitate the requirements of the motel. Guests have access to the exterior via balconies to enjoy the picturesque views of the suburban landscape. Aesthetically it is charming and architecturally it is a well-designed scheme. Unlike other projects of this nature this design is humble to its natural context.

Photography by Eresh Weerasuriya

Residence at Mahara

This residence adheres to environmental and microclimatic philosophies to enhance the architectural design. Climatic effects, construction materials, and techniques have been carefully designed to provide desirable conditions, while simultaneously enriching the architectural language. The residence is located in Mahara close to the Kandy to Colombo road. Designed on 26 perches, the west-facing site overlooks the main road. The methodology adapted for the design transforms microclimatic principles into a design tool that adds depth to the architecture and aesthetics, to create a home that performs adequately for the tropical climate.

From the gate, the view of this secluded structure is a composition of rhythm and textures that uses vernacular architecture principles but interprets a modern language. This concept is evident throughout the design to create a cohesive architectural dialect. The plan of the house is articulated along a central axis, which consists of the front veranda, living area, and the swimming pool. The additional spaces are connected along this defined hearth that also helps separate private and public spaces. From this core, spaces overlook one another creating visual corridors between the interior and exterior to reinvigorate living areas, inform, and formulate adjacent spaces.

Photography by Eresh Weerasuriya

Tom de Paor

Knockbawn Lodge
Mill Road
Greystones
Co. Wicklow
Ireland

www.depaor.com
office@depaor.com
+353 (1) 2557022

In the understanding that all practice is a form of research, our office maintains an open view of the permissible in terms of space, construction, scale, and permanence in the pursuit of an architecture that communicates to the user. Tom de Paor was born in London 1967 and graduated from the School of Architecture at University College Dublin in 1991. He has since taught and lectured widely in Ireland and overseas and received numerous national and international acknowledgements for his built work. He represented Ireland at the Venice Biennale of 2000, 2006, 2008, and in 2010 co-curated the national pavilion. He also designed a pavilion at the invitation of the Biennale International in that year. His publications include *N3* (Haus, 2001), *Circle Book*, *Reservoir*, *Of* and *previous*, *next*.

dePa

or

The Irish Pavilion

The inaugural presentation of Ireland at the *Seventh International Architecture Biennale (2000)* was made of land from Ireland, shipped to Venice, erected as the national pavilion and reused afterwards as compost for the Giardini at Venice.

Almost a sixth of the surface of the island of Ireland is bog. Vast tracts of this ground are harrowed annually to be raked, sun-dried, compressed, extruded, baled, and sold as fuel through local stores nationwide, a process which drains and destroys these landscapes and habitats. Some 40,224 of these peat briquettes, bound in orange polypropylene into 1,676 bales weighing 21 tons (19,050 kilograms) in total, were donated by the state and shipped and hauled 11,180 miles (1,800 kilometers) to Venice for use as the national pavilion. Sited on the promontory of land and dry dock between the Arsenale and Bacini and built on and of borrowed ground, the construction, its material, and its transport were consciously unsustainable.

The structure was corbelled in the manner and scale of early medieval Irish stone churches, familiar in the landscape and national iconography. The north and south elevations were stacked, and the baled extrusion profile allowed pinholes of light through the riven surface. Resultant from the extrusion process, the surfaces to the east and west appear polished. Corbeled in courses of three bales, rodded through these same holes and threaded between plywood fins, the corbels are able to disengage and complete the shifting stack of turf. We tied these buried timber buttresses back onto a base plate of plywood and they were anchored in place by the self-weight of the land.

Sited at Bacini in the Arsenale, the mud hut formed the letter N in plan and, oriented north, made an inhabitable, if temporary, north point in the archipelago. Two double-corbeled passages led to a fragrant top-lit chamber.

Typologically an oratory, only a few visitors at a time could be accommodated in its earthen room. In the quiet interior, there was a south-facing bank to sit on, the sky framed in the imported ground. Nine picture cards were presented on an embedded rubber cast for collection.

At the close of the Biennale, the Commune di Venezia accepted the international donation and demolished the structure, reusing the material as compost for the Venice Giardini.

Photography by Denis Mortell

Pálás (Picture Palace)

We designed an infill public building comprising bar, tearoom, and three cinemas in the footprint of a Georgian house and garden in inner city Galway, Ireland. The building was designed as a device for viewing, the architecture offering multiple overlapping vantage points, across the section and back into the town, orchestrated in the communal pleasure of cinema.

Located at the neck of land where the Corrib River rushes to the Atlantic and the city's docks, the building makes a gateway to the historic center and develops the municipal planning strategy to forge the cultural life of the small city eastwards.

It looks back to the public house tradition of the musical hall, the theater, and early cinema integrated with the urban grain in all its complexity. Like the towerhouse, the husks of which remain nearby consumed in the urban fabric, we completed the building as a shell and core that awaits inhabitation—leased for 25 years, as a cinema, bar, tearoom, and kiosk. The procurement strategy of this publicly funded building decided its construction.

Our architecture understands both permanence and change, from the austerity of the servant concrete to the luxury of the draped red curtain of the auditorium. This is rationalized and exploited in the layered construction—the detail and logic of the linings following the order of installation, relative longevity, and sustainability.

Consequently the structural envelope is dressed only where internally appropriate for acoustic, thermal, or light control. These depend on function and vary to liberate the inside of the concrete shell to be the inverse of the outside and spatially ambiguous like the ruin or a piece of town. The weather enters the building. A sliver of south light between the building establishes the foyer on the street and a little public space for the city.

We built this scheme in the plot of a demolished house and its garden. Its early 19th-century façade, a double composition around an arched carriageway, was considered important and remade to support the existing modest streetscape, the corner reinvented as gatehouse before the towerhouse of the cinema.

Typologically a towerhouse, the above-ground strategy of the cinema was to extrude the site boundaries vertically as massive concrete curtain walls, within which a thickened intermural service layer was wrapped around a stack of orthogonal rooms, each as large as possible on the limited footprint. The largest room is below ground within the plinth. We clad the ramped plinth in local limestone to become a rusticated base of dry stone walling integrated with the gardens and yards of the adjoining houses.

As you entered under a rebuilt limestone arch to a double-height hall, the stepped soffit of stairs overhead compresses the foyer, a top-lit outside place pinched in the tall thin gap formed between the building and an extension of the surrounding lanes. This space is served from a box office kiosk. A back staircase gives access to the piano nobile bar, again covered but external, while inside, the staircases are unheated and uninsulated before the completely tempered environment of the cinema rooms.

The building is a nine-story concrete monolith cast in a secant piled and tanked excavation. Rock anchors in conjunction with the mass of building resist floatation. Transfer slabs at ground and first floor level cantilever as the structural walls transfer due to their vertical discontinuity. A steel-frame tension and compression ring forms the pyramid roof of natural slate. The building is topped by a galvanized weather vane, which contributes to the skyline and locates the building in the city.

The concrete walls are shotblasted inside and out with chamfered rises and leading edges polished to reinforce the weight and homogeneity of the structure. Selected floors, ceilings, and façades are finished for tactility or light reflection.

Extensive use is made of the large areas of exposed thermal mass in the common areas to provide a sustainable civic building while harnessing passive solar energy and reducing active energy requirements. A night ventilation strategy allows the building to be pre-cooled for warm summer days. The individual theaters have heat recovery and carbon dioxide sensing for automated reduction of fresh airflows during periods of low occupancy. Only these three rooms have mechanical ventilation. The bar, tearoom, kiosk, and offices are all naturally cross-ventilated with the benefit of a five-story solar chimney on the south elevation to the bar, a triangular void which backlights the adjacent acoustic lobbies. The common circulation areas are uninsulated and unheated. We fitted them with low-energy bespoke lighting and they are naturally ventilated.

As a towerhouse, the services are stacked around its perimeter six-story concrete curtain walls, serving a tempered room at each level. The complexity of the circulation is generated between the servant and served spaces, as the building's plan mitigated the irregularities of the site to produce a dynamic townscape between rooms of different character.

The windows of the Pálás are oversized sapele casements; modeled like picture frames, hung on the interior, they pivot horizontally with integrated trickle ventilation. Proportioned in the Academy ratio, they are thought of as lenses, the interior of the double-glazed unit gelled with colored resin to designs by Patrick Scott (1921 to 2014), to infuse the interiors with color by day, and by night to act as lanterns to the town. They filter the hinterland.

The cinema relocates the Claddagh Palace, demolished in 1995, a decorated shed from 1949 originally called the Estoria, where the remarkable cultural cinema legacy of this small city developed. On the first floor landing, one window frames the view, across the slot in the building, of a cast of a local 17th-century bas-relief of Estoria, the maiden of the stars, incorporated into the formwork as a dedication.

The clear glass in the vertical slot provides views on three levels to overlook the foyer, the crossroads beyond, and to a glimpse of the River Corrib down the street. One staircase is concrete and treated as a stepped streetscape, while the other is steel, bolted together to span the triple-height room, as an object in the space.

Cast or removed lettering integrate the name of the building with its structure. At the gable to Dock Road, the letters were cut through to form the fresh air intake to the below-ground cinema, while on the flanks of the building they are cast in deep relief to amplify blue neon lettering on the scale of the city. The diacritics of the building's name appear in the windows to the staircase behind.

Photography by Anna Hofheinz and Peter Maybury

Clontarf

A municipal service building replaced the obsolete above-ground accommodation at Clontarf in Dublin. To upgrade utilities, working conditions, and amenities, the superstructure was demolished while maintaining below-ground waste-water services in operation. A new depot, pump house, and substation were cast in concrete on the existing foundations in the made-up ground. The public space of the village was also remade.

On reclaimed ground, on the edge between the north Dublin village and the estuary, the building houses the autonomous departments of parks, waste-water management, and electrical supply. Cast in concrete from a chamfered plinth on new ground, beams cantilevered from the original buildings deep below the ground structure, existing sumps, pump pits, and service gantry set out the building.

Typologically a shed, charged with the environmental services of water, power, and landscape maintenance, the form of the new building was developed as a necessarily robust closed urban object, visually dynamic to the curvature of the coast road, a building in the round to the existing linear park, while appearing axially downhill as a public building for the village before the sea. Although strictly functional, it is also picturesque.

We cast the walls in fair-faced concrete internally and clad them externally in a copper rain screen, which integrates the service access doors, with each department and function intimated on a separate façade. The continuous skin is treated with a pigmented sacrificial wax to control vandalism. The differential weathering of this surface exposed to town and sea at different orientations influenced our modeling of the enclosure. The photographs taken 10 years after construction demonstrate this.

The rain screen is eroded to the east façade to expose the concrete, which is canted, scabbled, and shotblasted to stain and weather between the existing trees. These uplit walls and slab combine to make a small place to play or gather in the shadow of the building, illuminated by three bespoke light standards. The building discharges the collected rainwater over the back-up generator extractor grill, to stain the shotblasted concrete shell and puddle its plinth.

Internally the building is matter of fact; exposed services are on an exposed structure for low maintenance and accessibility. A twisted beam over the back-to-back washrooms of each works department provides natural cross-ventilation. This strategy organizes the arrangement of the building on the most efficient footprint and maintains existing trees. The fresh air traverses the interdepartmental zone of the works to puncture the pre-patinated copper shingle rain screen twice—the ventilation is expressed through an untreated iroko grill. Natural light for periodic pump inspection is provided by a clerestory east light, which we applied to the concrete shell.

We rationalized existing car parking to prioritize the pedestrian, by planting 32 red maple trees for a roadside avenue—one tree outside each house, which gather to make a canopy over a small place for congregation and form a kind of signpost to the village for coast road traffic.

We developed a circular unit in three precast concrete sections into a triangular planter to allow tree planting over existing services. The joints were ramified to develop the ground plane with expansion joints in the in situ brushed cast concrete pavement, laid for drainage. We populated it further with rolled galvanized steel light standards and bicycle stands developed for Vernon Avenue, Clontarf.

Photography by Dennis Gilbert (opposite and above left); Peter Maybury (above right)

Christine Biesel
Principal

Wilhelm Eder
Principal

Valbergjet 11
4006 Stavanger
Norway

www.ederbiesel.no
post@ederbiesel.no

Eder Biesel Arkitekter was founded by Wilhelm Eder and Christine Biesel in 2008. The office has experience in projects ranging from large-scale urban planning, to small-sized family housing and interiors. By taking on different project types, our design response is also characterized by variation and diversity.

Our objective is to find serious architectural answers not only to core architectural tasks like office buildings, housing, student housing, hotels, and specialized buildings, but also to reinterpret less common architectural tasks like industrial and production buildings, such as warehouses or superstores.

Our core motivation for the design process is to improve and to surprise with the unexpected.

"The best is the enemy of the good," (Unknown).

"In this context we consider it necessary to question common ways and rules. There is no useful rule without an exception," (Thomas Fuller).

The perceived value of our work is not related to our practice but to each project, which aims to establish context-aware architecture.

Through our projects we continuously improve our expertize within many fields, in order to solve the current and changing requirements of sustainable architecture, in dialog with other experts and engineers.

Eder Biesel A

rkitekter

Fish Market in Bergen Climate Protected Marketplace

Bergen's fish market has had a long and important presence in Bergen. As a lively market place, it has left its mark on the city as much as the historic monuments that are significant tourist attraction in Bergen.

Our intention for this design competition was to carry on telling this story to future generations, hence the title *History Continued*. The market is located in the heart of Bergen, at the center of the World Heritage-listed area of Bryggen.

The position and shape of the new market hall maintain the sight lines from Almenning Road to the sea, well as nearby landmarks. Our aim was not to create an interior for the fish market but instead to provide climate protection for the marketplace so that that the limits between marketplace and market hall would disappear. The space is not divided. The floating volume is a roof and creates the protected part of the marketplace. The flexible glass façade provides shelter from wind and inclement weather. The façade can be opened on warm summer days. Market activities inside connect to those outside in a way that creates one homogenous marketplace and market.

Adjacent to the public section of the market hall is a three-story service entity. It houses storage space, cold storage and freezer space, an area for ice production, wardrobes and restrooms, as well other technical units. The market hall hosts the tourist information and the Norwegian Sea Food Centre on its first floor. The tourist information is also a public space with panoramic views towards Bergen's historic center.

The building's shape and the façade complement the historic context of this World Heritage-listed area. Even though the market has evolved as a new form, the choice of colors and materials emanate from the area's historic substance.

Historic origins are recomposed and reinterpreted with the use of glass. Wooden panels in colors typical for Bergen—ocher, dark red and white—change in rhythm and density, rendering a lively façade. From the front, the façade represents transparency and accessibility. This sense of accessibility and proximity determines a "modern marketplace" image both inside and out. The market's activity seems part of a staged scene. By night, the building structure appears to be beaming with light. The fish market in Bergen is a marketplace and center of city life, as well as a flexible open space protected from the changing climate.

Photography by Norbert Miguletz

Sandnes Sparebank Office Building

The new office building Sandnes Sparebank is placed like a piece of furniture in an anonymous industrial and car-oriented estate between monotonous square blocks. Construction processes in this area are characterized by a short time schedule and economic efficiency.

The design had to use efficient construction methods while simultaneously creating a symbol for Sandnes Sparebank. The concept is based on the main idea that the load-bearing prefabricated concrete elements are intended to define the building's character the lattice design is inspired by Austrian wafer biscuits called Manner.

Office areas are based on modules to ensure maximum flexibility. A compact building form with an atrium buffer zone and the use of the concrete elements as thermal mass contribute to reducing energy demand.

Photography by Øystein Thorvaldsen (left); Norbert Miguletz (below)

Troll Office Building

This office building, known as Troll, is located in Jåttåvågen Hinna, a former industrial area. The name Troll originates from a large concrete platform for offshore industry that was formerly built here. In contrast to the oil and gas industry's past history, one of the core priorities for this project's design was to achieve sustainability, high environmental standards and energy efficiency. Troll has been built as a passive house and an energy class A building (Norsk Standard). The project also focuses on reducing greenhouse gas emissions through the use of recycled materials and products.

Troll is designed as a compact volume. To save daylight in all building areas, a division runs throughout the building. This division also signifies the building's entrance. Galvanized steels is the main material used for the façade, with colors used in the reveal of the windows, following a strict pattern. The look and the reflection from the steel façade changes as one moves around the building. The colors only become visible when looking sidelong at the façade. The interacting design of form, pattern, and color creates variation and curiosity.

Photography by Norbert Miguletz

Rajesh Renganathan
Partner

Iype Chacko
Partner

Thekaekara House,
Banaswadi Road, 10th Main
2nd G Cross, Kalyan Nagar
Bangalore 560043
Karnataka
India

www.flyingelephant.in
rajesh@flyingelephant.in
+91 (98) 4535 7927

A common theme underlying our work combines experience of "nature" with a "sense of urbanity." The architectural sensibility I employ reveals an engagement with and play of opposites: old and new, nature and artifice, organic and crystalline, rough and polished.

We make a conscious attempt to engage "top down" and "bottom up" approaches in program interpretation and design process, working simultaneously from an overview of the "whole" as well as practical user-driven parameters of the "parts."

The choice of the composite construction system is yet another consistent marker, drawing as it does from a range of low cost vernacular, urban popular, and high-brow references without discrimination.

A strong modernist sensibility is evident at the core of our projects—a rational cerebral drive is very much in charge and the logic of structure, function, and material is earnestly followed through. But it is a modernism brought down to earth and rooted in the contingencies of everyday life.

Flying Elephan

t Studio

Healthcare Center

The project explores the inherent programmatic split of core clinic functions, and waiting areas in a primary healthcare center. The more expensive conventional core structure, built to higher medical specifications, accomplishes a weather tight, sterile, secure, and easy-to-maintain clinic building; while the innovative low-cost outer skin utilizes a material palette predisposed to available building skills in grass thatch and random rubble masonry. An additional outdoor landscape layer of varied cultivated terraces, medicinal plant patches, and orchards further helps mediate with the environment. Local vetiver grass thatch is used in the operable double skin, which creates a shaded climatic buffer around the clinic. This, in conjunction with excellent cross ventilation through the single banked structure, keeps the interior cool during severe tropical summers. The outer roof with prominent central gutter leading to a rainwater collection pond gives visual expression to water as a life giving element, also recalling folklore pertaining to the name of the place, Devara Outhu Pallam, meaning "God's spring village."

Photography by Manoj Sudhakaran

International Institute of Information Technology

The project is an extension to an existing academic building. A new innovative structure grows from the regular grid of existing columns; transforming its profile independent of the old grid, to form the raised roof of a winding internal street. The new façade is a natural outcome of the inner necessity of function, structure, and services combining with outer climatic response. This contrasts with the architectural character of the old building, which is graphically oriented. Rather than blend the two, old and new are juxtaposed, retaining their individual identities. The project was successfully completed with minimum alterations to the old building and minimum interference with existing service lines, while maintaining interior day lighting quality in the existing facility.

The proposed design allows maximum integration of architecture with the natural landscape, through the provision of peripheral views open to sky courts. Desirable green features of the distant landscape are consciously incorporated into the new building extension. A visual merger of old background and new foreground gives the illusion of landscape within, seamlessly extending to infinity. The perforated roof of the street incorporates circular skylight vents which provide a natural dappled light quality to the interior while allowing the building to breathe. An internal street runs past landscaped peripheral courtyards, opening up horizontal vistas to distant greenery and sky.

Photography by Manoj Sudhakaran

2DC Architects Residence

The image captures the layered "screen wall," which contains all the elements of the conventional "window," including security grills, insect mesh screens, glass, and curtains. The "skin" elements of a conventional urban dwelling are transformed into a new architectural arrangement to capture and intensify the experience of unique changes in nature. The local granite slabs, which are hand split on site, span between the fabricated steel joists, which in turn are supported on the steel I-stanchions and the brick walls. The sliding insect mesh screens in padauk frames slide up or down between I-sections as needed. The innermost layer comprises rolling chick blinds, which can be raised or lowered to suit light conditions or requirements for visual privacy.

Photography by Rajesh Renganathan

David Freeland
Principal

Brennen Buck
Principal

724 S. Spring St, #1101
Los Angeles
CA 90014
United States

www.freelandbuck.com
info@freelandbuck.com
+1 (323) 540-4346

FreelandBuck is an architecture practice based in New York and Los Angeles, formed in 2009. Affiliated with Yale University and Southern California Institute of Architecture (SCI-Arc), our office focuses on the spatial, atmospheric, and structural potential of computational pattern.

Contemporary western culture revolves around individuality. Our largest commercial, social, and political structures have fragmented to cater to our individual choices and tastes.

Buildings, rarely the efforts of an individual, must bridge this gap, allowing for the intimate within the institution, the unique as well as the shared.

At FreelandBuck, this is our design philosophy and field of research: how architecture can sponsor individuality in the context of our connections to each other and collective institutions. In other fields, Peter Slotterdyk's Spheres, Graham Harman's Speculative Realism, and Manuel Delanda's Assemblage Theory are all attempts to negotiate the relationship between individuals to collectives. In architecture, our most striking recent innovation, complex digital form, is more often used to set buildings apart as iconic objects rather than to incorporate multiplicitous spaces and activities. In contrast to the smooth surfaces and continuous forms of the last decade, our work finds continuity in the cellular, the aggregate, and the patterned. We see architecture as both part of the urban field and emerging from it, forming distinct spaces with an inherent connection to the activity around them. Through the use of visual, spatial, and organizational pattern, we attempt to produce new modes of synthesis and coherence, designing buildings and spaces that do not subsume their constituent elements but render their interdependence explicit.

Freelar

On the scale of installations and interiors such as our "Slipstream" installation in New York or our recent installation for Nike in San Francisco, we orchestrate material and structure in dense visual textures, which form a polarized field of potential activity and interaction. Rather than graphic, these textures perform spatially, shifting as the viewer moves through space and thus breaking down large interiors into distinct zones—of color, light, and texture.

At building scale, patterning, unlike smooth topological surfaces, collects disparate fragments within a continuous system while allowing for moments of inflection: differentiation, distinction, or dilution. For our Istanbul Disaster Center proposal and our winning design for the Kunshan Phoenix Mall in China, patterned and packed organizations allow distinct environments to bleed together, setting each room or zone in intimate proximity with its neighbors and maintaining a cohesive identity for the building as a whole. The result, we hope, are environments and structures that are both distinct from and interconnected with the world around them.

Our design thesis also serves as a methodology. Our office is split between poles in Brooklyn and Los Angeles, interconnected by technology, internal discourse, and shared history, but with feet in different markets and cultural contexts. Our practice is intertwined with our teaching work at two very different institutions, Yale and SCI-Arc, and two distinct sets of collaborators, consultants, and contractors. Our practice is an ongoing search for cohesion between these separate geographic and cultural environments.

ndBuck

Earl's Gourmet Grub

An artisanal deli and gourmet market, this 968.8-square-foot (90-square-meter) tenant improvement is conceived of as an interior landscape; a variable and shifting space defined by a series of torque ceiling surfaces and light "scoops." The ceiling creates an airy, light-filled canopy with local intensities modulated by wood "baffles" of oscillating depth. These rhythmic undulations both subdivide the linear space into a series of spatial pockets and produce dynamic spatial continuity from front to back.

Architectural computation is generally promoted in relation to high-tech building systems and iconic towers. The design for this space is a test case in how computational architecture can enrich a small space's everyday use. The restaurant was designed with cutting-edge technology to fit an old-world sensibility inspired by its food. Torquing ceiling surfaces and inscribed digital patterns are combined with a rich material and color palette to evoke both technological refinement and the more rustic feel of alpine landscapes and Viennese cafés.

The undulating baffles along the west wall are thought of as a long continuous light fixture. Three long series of industrial fluorescent strip lights variably coincide with the oscillating baffles, creating alternately warm and less intense reflections. Sequenced switching allows the client to tune interior lighting to the time of day and intended mood.

Photography by Lawrence Anderson

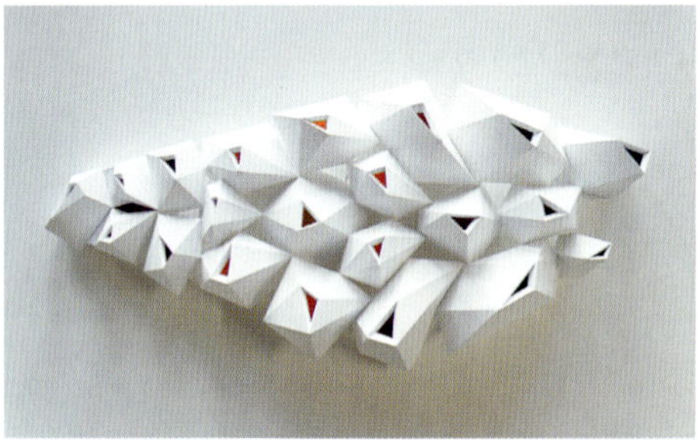

Disaster Information Center

Prompted by the expectation of a major earthquake in Turkey, the Istanbul Disaster Prevention and Information Center holds a series of disaster simulation galleries, from earthquakes to rainstorms to fires. We designed the building as a single structure, which expresses the multiplicitous atmospheres and climates it contains. Unlike a science museum, where large open halls are sufficient to contain furniture-scale exhibits, the disaster center is a continuous network of distinct spaces. Each exhibit relies on a unique atmosphere and climate supported by its own structural, mechanical, and electrical infrastructure. The exhibit rooms take the form of "boulders" with a single central space at the core surrounded by a ring of infrastructural poché and circulation space. The core of each boulder is a zone of intense atmosphere produced through dramatic effects of lighting, climate, and color, which leak out to infuse the more neutral lobby and reception spaces around them.

Yale Assembly One Pavilion

Designed, fabricated, and assembled by Brennan Buck's students at Yale University, the "Assembly One" pavilion is the younger, more carefree sister to Yale's building project—the 40-year-old tradition in which first-year students design and build a house. Designed as the information center for New Haven's summer International Festival of Arts and Ideas, we developed the pavilion with several characteristics in mind.

Dynamism: the structure is suited to a performance festival—solid and massive from one angle, lightweight and almost entirely porous from another, it alternately hides and reveals its contents.

Visually Transparency; constructed from thin aluminum sheets, the pavilion opens up on two sides for ventilation and security, focusing views toward the festival's main stage.

Visual Density; over 1,000 panels create shifting effects of reflection and color as visitors move around the pavilion, creating less of a timeless image of shelter than an unstable, engaging heart of the festival.

Photography by Chris Morgan

Yves Béhar
Founder and CEO

1401 16th Street
San Francisco
CA 94103
United States

www.fuseproject.com
info@fuseproject.com
+1 (415) 727-2501

I am a design entrepreneur who believes that holistic product, digital, and brand design are the cornerstones of any business. I am the founder and CEO of fuseproject, an integrated design and branding firm I established in 1999. I am also the Chief Creative Officer of Jawbone, winner of a 2010 IDSA Design of the Decade Award, and the co-founder of August, the next generation home entry system.

I believe that design should be a force for positive social and environmental change. My humanitarian work includes the One Laptop Per Child, which has given 2.5 million laptops to children in developing countries, and See Better to Learn Better, which distributes free corrective eyewear to schoolchildren in Mexico and California. For each of these I have been honored with the INDEX Award, making me the only designer to have received the award twice.

My collaborations with renowned partners such as Herman Miller, GE, Puma, PayPal, SodaStream, Samsung, Issey Miyake, Prada, Nivea, and many others have received international acclaim. My works are included in the permanent collections of museums worldwide, including MoMA, the San Francisco Museum of Modern Art, the Centre Pompidou, and the Art Institute of Chicago. I am a frequent speaker on design, sustainability, and entrepreneurialism; I have given talks at TED, the World Economic Forum in Davos, and the Clinton Global Initiative.

I am currently the artist trustee of the Board of Directors of the San Francisco Museum of Modern Art, have garnered over 200 awards, was named a Top 25 Visionary by *TIME* magazine, and was recently named "Most Influential Industrial Designer in the World" by Forbes.

roiect

XO Laptop

The OLPC XO Laptop (also dubbed the $100 laptop) is an Index: Award winning machine which I designed with children in mind from start to finish. The XO Laptop needed to be small and light for children, but sturdy and robust to ensure durability and to prevent damage from natural elements and years of use. Because the laptop is for children, the aesthetics needed to appeal to children.

Designed for intended use by millions of children, the XO Laptop needed to be resilient, easily portable, and energy efficient.

My design brings incredible value to the function, experience, and feel of this object, and specifically, a design that is expressive at the same time (like the fun rabbit ears). OLPC XO proved the real impact the OLPC mission will have for kids worldwide: a new dignity and pride to possess their own education, communication, and entertainment tool, making the XO a true industry game changer.

Photography courtesy of fuseproject

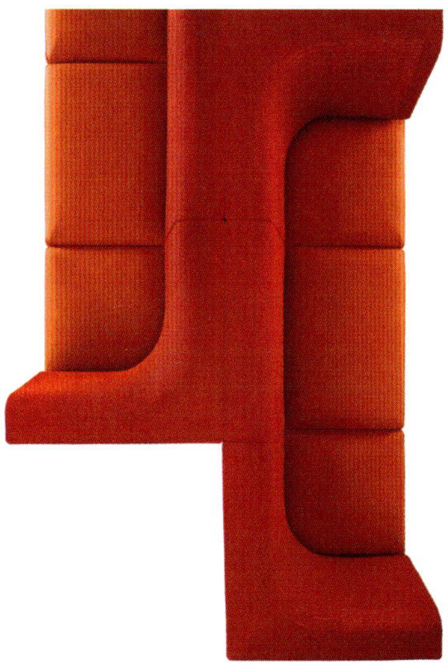

Socializing

An admirer of the legendary Herman Miller designers, Charles and Ray Eames, Yves Béhar declares: "I took up the phone one day and called the furniture company to suggest a working relationship." The result is a collection that brings new wisdom to the workplace. Béhar's Sayl chairs come in several iterations, as ergonomic task seating with or without arms and with a sled or stacking base. Béhar's systems solution, Public Office Landscape responds to the new flexible workspace preferred by startups and high tech organizations where an impromptu meeting of the minds is often the preferred way to hatch ideas. Béhar's "Social Chairs" interactive workstations and collaborative layouts help to support creative thinking.

Photography courtesy of Herman Miller Inc.

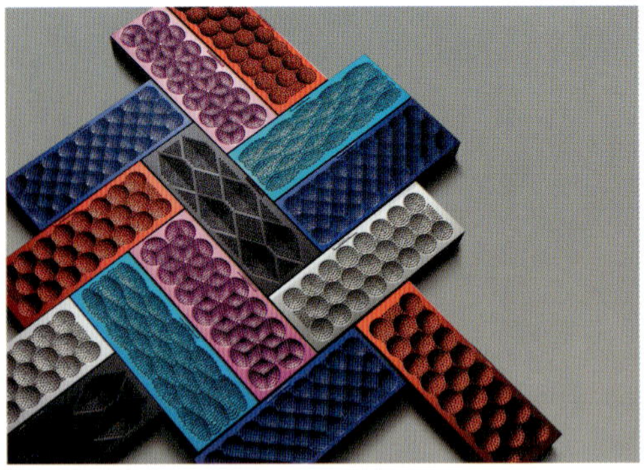

Mini Jambox

The Mini Jambox is the latest entry in the category-leading family of Jawbone speakers. When we first explored what the Mini Jambox could be, we dreamed of a very small and pocketable size, of an experience so simple and yet game changing, and of materials and processes so refined that they had previously been used only in top-end audio products.

Jawbone design goals are to seamlessly integrate technology and everyday life. The Mini Jambox is built on the foundation that life is constantly moving. With Mini Jambox you can pick up and bring your sound environment with you. We call it pocketable sound.

Photography courtesy of fuseproject

Clever Little Bag

We partnered with PUMA to redesign the shoe box into the Clever Little Bag. The cardboard structure is die cut from one flat piece of material and has no additional printing or assembly—thus it can be returned to the stream faster and more efficiently. The structure was created with four walls which taper in to allow for secured stacking, another important element left over from the original shoebox. The bag is nonwoven, which means less work and waste (it is stitched with heat). It protects the shoes from dust and dirt in the warehouse and during shipping. The "clever little bag" is an iconic brand element upon leaving the store as it replaces the plastic shopping bag, and can also be repurposed for creative reuse. The bag is made of nonwoven polyester consisting of polypropylene, and eventually is also recyclable. With our Clever Little Bag, Puma kicks off the next pivotal phase of its sustainability program. The tens of millions of shoes shipped in our bag will reduce water, energy, and diesel consumption on the manufacturing level alone by more than 60 percent per year. In other words:

Approximately 8,500 tons less paper are consumed, 20 million mega joules of electricity saved, 1 million liters less fuel oil used, and 1 million liters of water conserved. During transport 500,000 liters of diesel are saved and lastly, by replacing traditional shopping bags, the difference in weight will save almost 275 tons of plastic. That such a little bag can have such a big impact...you can see why we called it clever.

Photography courtesy of fuseproject

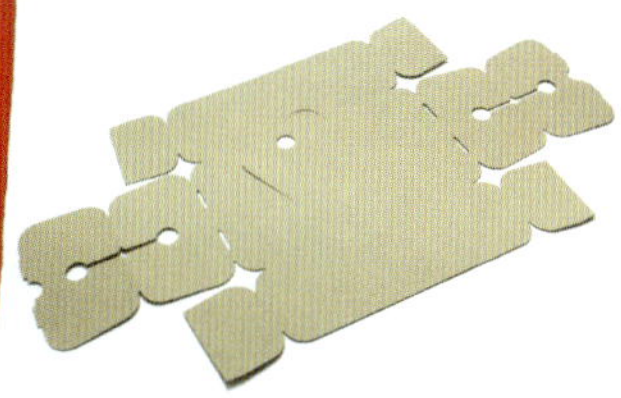

Pascal Grosfeld

Reduitlaan 39
Breda 4814DC
Netherlands

www.grosfeldvandervelde.nl
p.grosfeld@grosfeldvandervelde.nl
+31 (76) 5221811

Architects and their clients work in an environment in which people live, work, relax, and attend school. That is a task that brings with it considerable responsibility and one that we approach with a strong belief in serving society. The idea is not to build in honor of the designer, but to create suitable buildings for general use in a public environment. A sound building also meets wishes and demands at various levels without being affected by temporary trends. The best buildings combine clarity and peacefulness with uniqueness and appeal. We always seek out their essence and try to create simple and yet striking buildings. We believe in craftsmanship, good design, and properly constructed buildings, which have clearly received attention. The aim is also to use available, but often limited, resources as effectively and smartly as possible. Besides searching for the best solutions for functional demands and technical requirements, another of our explicit aims is to create the right perception, atmosphere, and character. The form is the result and not the goal. A cherished building lasts longest—that is the highest form of sustainability.

Grosfeld Velde Ar

We regard renovation and conversion as the most important task for the future. Because of its high degree of complexity we view it as one of the most interesting assignments imaginable. Each specific context requires an architect's utmost effort, as well as a solution that is unique in every respect.

We firmly believe in the necessity for sustainable conversions and renovations of existing building stock. In this regard we deliberately opt not for the strategy of conserving, but for effective and efficient reinterpretation and transformation. Existing building stock must be preserved for the future by transforming it respectfully, in accordance with the functional and technical requirements of our time. Having said that, we are not afraid to make modifications. But, we always do so on the basis of achieving the correct balance with a building's authentic character. A reinterpretation of the original, historic features is extremely important. The question is, when does transformation cause a building to lose its intrinsic quality and when is this quality still clearly preserved despite modifications? It is this search that typifies our vision of conversion and renovation.

Over and over again house designing offers us architects the opportunity to connect with the essence of our profession. More than any other architectural task, designing houses for private clients brings out the best in us. A good architectural design is possible only when we feel an intense, intimate relationship with the client. Although it sounds contradictory, the limits imposed give us the freedom to achieve a high degree of uniqueness.

A private house is the ultimate in architectural uniqueness. Its architecture exists only if there is a context and that is what we believe is the beauty of our profession. The context, in conjunction with a unique assignment, gives architects the opportunity to show architecture at its beautiful best.

van der

chitecten

House of Culture Garenspinnerij

The Koninklijke Goudsche Machinale Garenspinnerij N.V. building dates from the beginning of the twentieth century. This example of Dutch industrial heritage located on the Turfsingel in Gouda is characterized by plenty of light, distinctive brickwork, and a rigid concrete and steel construction.

The schedule of requirements detailed the conversion of the Garenspinnerij building into a cultural center to accommodate a variety of the city's cultural institutions. The intrinsic qualities of the building became clearly visible once it had been stripped back to its original and authentic shell. On the upper floor, the new design is literally kept separate from the existing building, making it possible to fulfill noise and comfort requirements. Opting for a different, abstract, but also unassuming and unequivocal architectural style throughout reinforces the authentic and scarred shell. At the same time, the new construction derives its identity from the existing building and represents a new phase in its history.

Photography by Rene de Wit

House L

The plot of House L is sited in the wooded environs of Oosterhout. The existing landscape, the orientation to the rural surroundings and planning conditions were the deciding factors in siting the dwelling at the rear of the plot.

The ground floor was raised above the current ground level, with large glass surfaces positioned to look out over the rural landscape and terraces. These run the length of the building with an unbroken roof surface on corbelling, all of which allow the outer space to be experienced to the full. Simultaneously, the large roof projections were designed to prevent too much light from entering.

The ground floor may be described as a single, open space where the massive volumes and project-oriented design of the interior provide a further connection with the outside. This floor accommodates the primary living functions, with all rooms flowing seamlessly into one another. The annex accommodates a number of ancillary functions, with the outdoor swimming pool housed in an extension. Both volumes are connected underground by a basement level in which the bedrooms and bathrooms are situated. Here too, the landscaping of the grounds and the large glass windows allow for outstanding views over the surrounding landscape.

Photography by Loes S (top and middle); Marcel Briaire (below)

Utrecht University Library

The design is based on the concept of creating a natural meeting place for education and research, for employees and students, and for the university and the city of Utrecht, in the heart of the Drift Cluster. This is a place where students and books take center stage. The core values for the identity and perception of the complex are accessible, transparent, inviting, representative, and restful. In addition to providing a functional and pragmatic solution for the program of requirements, our design accentuates the intrinsic qualities and spatial layout of the different buildings, each of which has its own strengths and complements the others. The materials, details, and color scheme have been specifically chosen to draw the buildings together to create a unified whole.

Most of the study areas are situated at the front of the library and the collection is at right angles to the front of the building in an open arrangement. This makes the building light and transparent and encourages interaction. It creates a studious, attractive work environment where students can walk in and around the collection at their leisure and where there is space to meet for informal contact.

Photography by Rene de Wit

Bijoy Ramachandran
Partner

Sunitha Kondur
Partner

9/7 Madras Bank Road
Bangalore 560001
Karnataka
India

www.hundredhands.com
bijoy@hundredhands.com
+91 (98) 4569 2652

Our studio's approach is grounded in the search for contextually appropriate solutions. We give special emphasis to the response to climate, the existing scale and character of the context, appropriate use of materials and construction techniques, and the development of the project aesthetic as a result of these specific conditions.

In working with the locally available materials, recognizing local skills, and directly responding to the existing conditions, we are continuing an age old practice of being prudent and responsible. The current predilection of talking about a green architecture or sustainable architecture seems redundant. We are always thinking of an all-inclusive practice—an architecture without adjectives.

Through our work we are trying to address concerns which are larger than the briefs given to us by clients—we want to engage with the real opportunities India provides us—the chance to work with sophisticated craftspeople and specialized fabricators, the chance to reinvent mechanical systems to address building techniques and sustainability issues, and the chance to look at buildings beyond building systems and in terms of principles.

Hundred

lhands

Alila

Exterior from northeast showing the organization of the program—common amenities on the ground and fifth floors, hotel in between and apartments on the upper floors (left); Ground floor has an open verandah (above top); Interiors and architecture working together to help clarify structure and organization (above middle); Hotel room interior—materials inserted into and clearly articulated from the structural shell (above bottom)

Photography by Nathan Willock (left and above top); Mallikarjun Katakol (above middle); Adri Berger (above bottom)

Nua

Entry façade—the material palette is expressed (right); The courtyard revealed on entry to the gallery (middle); Rooms and verandahs open onto the courtyard (bottom)

Photography by Mallikarjun Katakol

Taaqademy

Light from the long window on the western side casts patterned shadows on the brick walls (above); View on entry, curvilinear shapes—housing recording/jam rooms—within a simple orthogonal enclosure (left); A glimpse of the music room within (below left)

Photography by Mallikarjun Katakol (above and below left); Neha Bajaj (left)

Janne Kyttanen
Principal

Hobbemakade 85hs
1071 XP Amsterdam
Noord Holland
Netherlands

www.jannekyttanen.com
press@jannekyttanen.com
+31 647902740

When I first discovered 3D printing I immediately saw everything around me in wireframe and I was completely overwhelmed by the possibilities. I saw all of the potential of the technology and I didn't see any reason to go into a workshop and cut wood, or bend metal, or weld—I immediately submersed myself into the computer and I never let go.

I realized that every object I looked at could be stored virtually and produced anywhere with very little environmental impact. I have been working with 3D printing for over 15 years and from the very beginning I envisioned a future in which there would be a profound change in the way objects are manufactured. The benefits of working with this technology are unprecedented when compared to existing manual labour-intensive ways of making: speed, detail, and complexity.

Janne Kyttan

From the beginning I embraced augmented reality. The plan was to create a world in which people could download data into their living rooms, view this data through stereoscopic glasses, haptically adjust the geometries when needed, and then print in 3D what they needed. At this time I was more interested in becoming the architect of a totally new industry, than focusing on designing products. I created for myself a dream world in which I could roam, however, I pleased with my creative ideas.

My creative process has never really involved thinking about and executing specific concepts. New ideas come to me; I don't try to force them. I intuitively steer myself towards new things all the time and I prefer never to make the same thing twice. I have a tendency not to look back and to reminisce about the past. My curiosity pushes me forward, giving momentum to my work.

hen BV

Laser Sintered Light Designs

I started designing lights in 1998 and took the idea of creating a collection of 3D printed lights to Materialise. They were excited by the idea and invested in creating a whole new department, initially focused on making lights.

Over the years I have developed a full portfolio of light designs, taking inspiration from organic forms. These lighting features are all 3D-printed—the material that they are created from (nylon) has a special quality, allowing light to shine through the structure and cast light onto surrounding surfaces at the same time. The result is a glowing, organic form and ambience-creating light and shadows.

Photography by 3D Systems (opposite, middle left, below left); Materialise (top left)

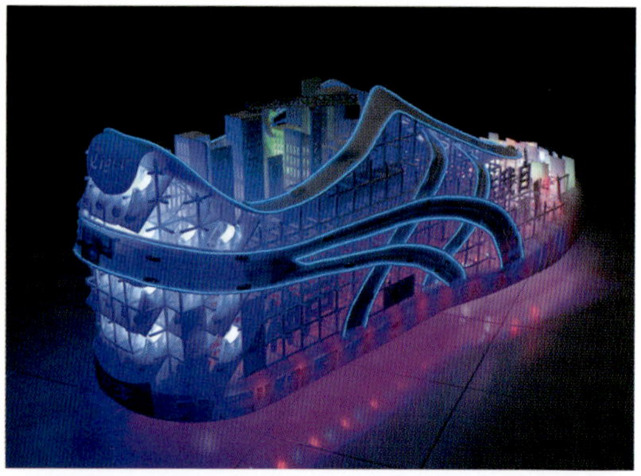

Electric Light Shoe

In 2008 I created the Electric Light Shoe for the Onitsuka Tiger "Electric Tiger Land" campaign. These images show the 1-meter-long (3-foot) Electric Light Shoe with illumination.

This design was central in the brand campaign and features a miniature microcosm that mimics the hustle and bustle of Japan's urban centers.

In addition to the feature design, I created in-store displays and small accessories such as USB keys and metal key rings, squeezing the intricate detail of the full design into a handheld form.

Photography by Amsterdam Worldwide

Lost Luggage: mixed 3D printing technologies

Imagine if everything we needed was scaled down into a digital file and could be reproduced at any location; how would this change our relationship to physical objects?

Imagine that design is just data and products can travel through the internet as code, produced on demand at any location. If products and designs are distributed in the same way as music and images travel through the internet today, how would our perception of those objects change? Could we bring production closer and closer to the consumer, regardless of their location and economic status? And would this localization of manufacturing eliminate the need for mass production?

What if luggage was obsolete? How would this change our perception of travel? Send your luggage in an email, travel the world unencumbered, and arrive at your destination to find your luggage waiting for you. This future is already within our reach.

Photography by Adrian Woods & Gidi van Maarseveen © 3D Systems

Alice Kimm
Principal

821 Traction Avenue
Suite 107
Los Angeles
CA 90013
United States

www.jfaknet
akimm@jfak.net
+1 (213) 253-4740

Our work at John Friedman Alice Kimm Architects begins with the premise that architecture dramatically affects people's lives. We exploit this potential to create new types of social spaces that encourage creative interaction, sustain the environment, and catalyze positive change for our users and their surrounding communities.

Our distinct formal landscapes are the result of an inquisitive design process that mines the possibilities and complexities of our contemporary, heterogeneous society. Every project we undertake is committed to expressing the aspirations and identity of the entity it represents through productive dialog with its physical and cultural surroundings. Light-filled interior spaces are organized to create meaningful adjacencies and unexpected views, as well as to encourage a range of activities—from individual contemplation to communal interaction and creative teamwork. Exterior form is both reflective of this activity and sculpted to reflect the exigencies and influences of its surroundings.

The size, type, or budget of a project is not important to us. Rather, we take on projects that involve a meaningful design challenge and the promise of developing an innovative solution. This effort often involves the utilization of new types of performative materials in concert with current design software and fabrication technologies. For us, new digital techniques are not an end in themselves; they are tools to explore new types of efficiencies, forms, spaces, and branding possibilities. It is, therefore, always interesting to see what happens when age-old techniques and tools are combined with contemporary ones, and are indistinguishably woven together through intuition, imagination, and the resources at hand.

John Friedm
Alice Kimm

Ultimately, what is produced is, to us, an embodiment of culture—and as technology increasingly takes over our lives and our work, we see culture adapting and expanding to encompass all of what comprises our productive and creative lives. As our work evolves, interdisciplinarity and the cross-pollination of different cultural (and technological as a subset of cultural) influences become more and more our inspiration and guide.

Finally, as a matter of ethical and environmental urgency, we apply a wide range of sustainable strategies, whether they involve the most sophisticated engineering systems, simple passive techniques, or the choice of the most responsible materials. Casting our net as wide as possible, we collaborate with an ever-expanding circle of artists, engineers, scientists, builders, and other specialists. Whether these individuals are our clients, consultants, or advisers, they become partners in a shared vision—one of creating integrated and sustainable environments that inspire their users and improve their lives.

nan

Architects

La Kretz Innovation Campus

Currently under construction, the La Kretz Innovation Campus is a 64,500-square-foot (6,000-square-meter) transformation of an existing one story masonry warehouse. It will house the Los Angeles Cleantech Incubator, as well as Los Angeles Department of Water and Power offices and demonstration labs. The Incubator will contain open as well as private offices and conference rooms, open areas for interactive communication, prototyping labs, boardrooms, meeting rooms, and other types of flexibly configured spaces. Designed and engineered for LEED Platinum certification, this facility will become a new type of business hub in what was formerly a manufacturing district, and will set a new standard of how adaptive reuse can rejuvenate aging areas of the city.

Renderings by John Friedman Alice Kimm Architects

Koreatown Gateway

The Koreatown Gateway will be located at a major intersection in the Koreatown neighborhood of Los Angeles. This cultural marker will embody the strength of the ties between Los Angeles and Korea, and will express the commitment to innovation and the advancement of a vibrant cultural life which is shared by both places. The project is comprised of two cantilevering steel beams between which is draped a cable net canopy enlivened with an LED array. The work of local artists, such as that of Yunhee Min as shown in one image on these pages, as well as other types of imagery, will be interchangeably programmed to create a dynamic work of public art.

Rendering by John Friedman Alice Kimm Architects

Graduate Aerospace Laboratories at the California Institute of Technology

We undertook this 18,000-square-foot (1,675-square-meter) renovation of the historical 1928 Guggenheim Laboratory at Caltech on behalf of the institution's Graduate Aerospace Labs (GALCIT). The program includes a new lobby, laboratories, exhibition areas, conference room, offices, and interactive spaces.

The design utilizes current design software and fabrication techniques to create an environment that reflects the innovation occurring within the department, and to create an updated new identity to attract students and faculty from across the world.

Photography by Benny Chan, Fotoworks

Sharon Johnston
Partner

Mark Lee
Principal

1545 Pontius Avenue
Los Angeles
California 90025
United States

www.johnstonmarklee.com
sharonj@johnstonmarklee.com
+1 (310) 442-4886

Recognizing 15 years of distinguished practice, Johnston Marklee received the AIA Emerging Practice Award, Los Angeles Chapter, in 2013. Our process investigates issues of design, structure, and technology to create unique and vital works of architecture.

Principals Sharon Johnston and Mark Lee have honed their approach to projects, varying in scale from master plans to contemporary buildings and temporary installations, to distill the inherent complexity of each project into coherent and singular solutions. Typologically, we have a substantial architectural portfolio including residential, institutional, retail, commercial, and hospitality projects—with a particular focus on the arts and an inherent commitment to employing innovative building practices.

As a practice we maintain long-term collaborations with art and cultural institutions, commercial developers, and private clients. Our current projects include the Menil Drawing Institute (MDI) in Houston, Texas—a new 32,000-square-foot (3,000-square-meter) building dedicated to the exhibition, study and conservation of the Menil Collection's extensive drawing collection (Start of Construction: 2015); the Grand Traiano Art Complex in Grottaferrata, Italy, and the Poggio Golo winery and Artist-in-Residence studios in Montepulciano, Italy, for DEPART Foundation; the recently

Johnston

completed 5,000-square-foot (450-square-meter) gallery building for the West Bund Bienniale in Shanghai; and META, a community arts center in Penco, Chile, for the Chilean government. Johnston Marklee has also formed long-term partnerships with the Hammer Museum to collaborate on exhibition design and planning, including shows for James Welling, Robert Heinecken, and a group exhibition on Frottage.

Our work is recognized internationally for its engagement with contemporary art practices while being deeply rooted in the history and foundations of the discipline. Our firm draws upon an extensive network of collaborators in related fields, including contemporary artists, graphic designers, writers, and photographers to broaden the breadth of design research.

Artist collaborators include Ai Wei Wei, Walead Beshty, Jeff Elrod, Luisa Lambri, Takeshi Murakami, Jack Pierson, and Sterling Ruby. We ascribe to the model of collaboration in which the expertise of joining disciplines is sharpened, rather than blurred, maintaining permeable boundaries for greater results. A forthcoming book entitled *House is a House is a House is a House* is structured around five artists' interpretations of the work of our firm. The artists include Livia Corona, Veronika Kellndorfer, Luisa Lambri, Jack Pierson, Walead Beshty, and James Welling, who frame our projects from the perspective of their own work.

Further expanding the firm's involvement with the design community is our participation in architectural education. Both of us have held faculty positions in the Department of Architecture and Urban Design at the University of California, Los Angeles, and have lectured and exhibited extensively throughout the United States and Europe. We were Cullinan Visiting Professors at Rice University during the spring 2013 semester and have been teaching a multiyear studio at the Harvard Graduate School of Design since late 2013. Our firm considers academic exchange as critical for the firm's work and as an essential contribution to the development of future architects.

Vault House

East view (opposite); Exterior (above); Living room looking west (right); Living room looking east (below right)

Photography by Eric Staudenmaier

View House

East view (above); North view (right); Stair/kitchen (far right)

Photography by Gustavo Frittegoto

Hill House

Southwest view (above left); Living area (above right); Exterior (left)

Photography by Eric Staudenmaier

Kashef Chowdhury
Principal

House 56, Road 5A
Dhanmondi
Dhaka 1209
Bangladesh

www.kashefchowdhury-urbana.com
kashef.chowdhury@gmail.com
+880 (2) 967-1515

The work of Kashef Chowdhury/URBANA finds root in history, with strong emphasis on climate, materials, and context—both natural and human. All works in the studio are given extended time for research, in order to reach a level of innovation and original expression. Works range from the conversion of a ship and low cost raised settlements in remote river islands to training centres, mosques, art galleries, museums, cyclone shelters, residences and multifamily housing, resorts, corporate head offices, and large masterplans.

Compelled to take a pragmatic approach in light of economic and other constraints facing my commissions, I pursue "an architecture of responsibility," where optimization of resources, efficiency, and clarity of design and an aesthetic sympathetic to historical inheritance and societal lineage assume prime focus. The resultant work is not an architecture of the minimal, but of the essential.

I take an active interest in art, photography, and music, which has lent inspiration to my work.

Kashef Chowdhury

/URBANA

Chandgaon Mosque

The forecourt (above right); View from northeast, from the approach road (right); The dome and the circular opening at night (far right)

Photography by Kashef Chowdhury

Friendship Centre

Green roofs of sunken volumes of the centre merge with the rural landscape (top); Spaces are woven out of pavilions, pools, and gardens (middle); Afternoon light bathes the dining hall, finished and furnished to simple detail and specification (bottom)

Photography by Hélène Binet

Museum of Independence and Independence Monument of Bangladesh
In collaboration with Marina Tabassum.

Approach to plaza (above left); Museum interior (above right); Independence Monument, mirrored in the surrounding reflecting pool (right); The heart of the museum is a space for contemplation (opposite)

Photography by Kashef Chowdhury

Diébédo Francis Kéré
Founder and Principal

Arndtstrasse 34
10965 Berlin
Germany

www.kere-architecture.com
mail@kere-architecture.com
+49 (030) 78952391

As a native of Burkina Faso, I grew up in a region with many economic and climatic challenges and very few resources. When I was a child, I had to travel to another village to attend a school, which was poorly ventilated and badly lit on the inside. The experience of trying to study and learn in this hot and dark classroom affected me so much that when I began to study architecture and construction in Europe, I had the idea to build a better school in Gando, my home village.

When I began to make plans for the primary school, I hadn't even graduated from university. At the beginning, many of my peers and professors discredited my intention to build in Gando. I had no resources, no money, and little experience, but I had a strong idea and a lot of energy. I returned to Gando with the idea to build a school out of clay, a place where the term "architecture" simply did not exist.

How do you build or create a sustainable infrastructure without money or sophisticated technology? I use what is already there to my advantage: clay and community. This is how I began to make architecture. I use local resources and combine modern construction techniques with traditional building methods. Some people call me an expert in clay construction, but I see myself as something more. I am someone who can communicate complex ideas about construction with communities which sometimes can neither read nor write. For my people in Gando, I am like a bridge connecting them to new opportunities. To move forward, people need to be inspired: they need buildings that enhance their creativity and push them to take their future into their own hands. This is my philosophy.

Kéré Arc

hitecture

Gando School Library, Burkina Faso

Exterior view of library before wood façade elements were added (above); Casting the ceiling with locally produced earthenware pots (right); Day-lighting effect of ceiling with cast earthenware pots (below right)

Photography by Kéré Architecture

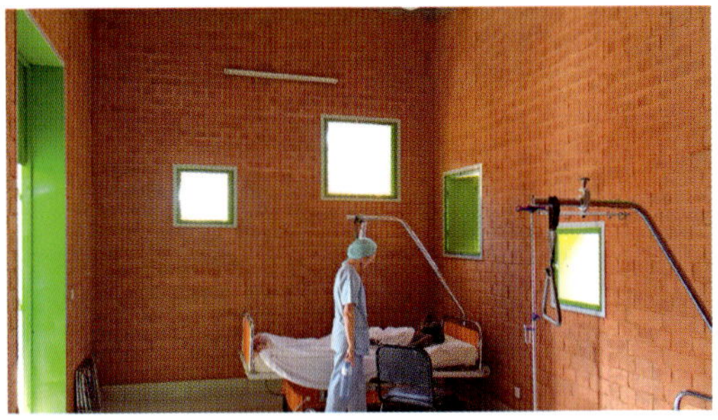

Léo Surgical Clinic and Medical Center, Burkina Faso

Exterior view of main outdoor corridor with clinic modules (top); Interior view of examination room (middle); Exterior view of clinic modules (bottom)

Photography by Kéré Architecture

Opera Village, Laongo, Burkina Faso

Interior view of Opera Village Clinic courtyard (left); Exterior bird's-eye view rendering (top); Exterior view of Opera Village School (middle); Exterior detail view of fenestration and façade of Opera Village Clinic (bottom)

Photography by Kéré Architecture

Koen Olthuis
Founder and Principal

Gen. Berenschotlaan 211
2283 JM Rijswijk
Zuid Holland
Netherlands

www.waterstudio.nl
info@waterstudio.nl
+31 (70) 39-442-34

In addition to climate change, the fabric of society that is composed of political, economic, and social threads is also changing. With each passing decade, it is becoming more and more difficult for us to predict what the needs of tomorrow's city will be. When a city's infrastructure no longer satisfies the community's needs, demolition and reconstruction are considered the common solution, even if the structures have technically not even come close to the end of their expected lifespan. By providing a resolution to many problems stemming from climate change and urbanization, water based developments are redefining urbanism.

Just as the elevator was essential to use the space above our cities, modern floating foundations can offer flexibility sought after by contemporary designers and urban planners, as well as an ample amount of space in the forms of rivers, lakes, canals, and coastal zones.

A floating development is used, re-modified, restructured, moved, and reused until the product's lifecycle has been completely exhausted, at which point it is recycled, and all waste associated with this process is appropriately discarded. The dynamic quality of floating developments and the associated system of production offer a new method of construction which is minimally invasive.

Koen O

Since most modern urban developments have a functional lifespan of approximately 50 to 80 years, a home or business owner is considered a consumer of a product with a limited lifecycle, which will eventually expire. Similar to how modern technology provides customization tools in order to organize and seamlessly integrate virtual amenities into a person's life, floating technology strives to make use of its adaptability to shorten response time and offer mobile urban components known as "City Apps." Adding a floating app to a city is similar to adding an application to a smart phone. City Apps are floating urban components with many different kinds of functions, which are added on water to an existing city's static grid. They are created off-site then moved into an area, providing instant solutions which respond to a society's changing needs. City Apps are easily moved and updated, which opens up the possibility of relocation, lease, or resale by transferring them elsewhere when and if there is no longer a need for them in their original location.

Floating technology may not be the answer to every problem, but water based urbanism is set to transform communities around the world. Traditional, static cities will become more dynamic. The foundations necessary to entertain innovative creations that will give shape to the future already exist. Changing the perception towards water based developments is crucial to finding opportunities that will provide greener, safer, and more flexible urban developments. An open mind will prepare cities for the global challenges yet to come by providing economic, political, and environmental freedom and prosperity.

The city of tomorrow is fast, flexible, conscious, and water based. It has the courage to think outside the box and not to rely on conventional ways that might prove insufficient to meet modern challenges.

lthuis

Wet Slums

One billion people live in slums and 30 percent of them are in wet slums. These people are the most vulnerable and neglected in upgrading programmes. By leasing floating functions in containers to entrepreneurs in slums, their quality of life can be improved. These so-called City Apps are flood-resistant, take up space only on water, not land, and can be moved to other locations and slums when situations change.

Rendering by Waterstudio.NL

Maldives 2.0

The Maldives are threatened by a sea level rise and lack of dry land. We developed the floating projects in a joint venture with the Maldives government, who wished to change their national profile from doomed climate refugees to climate innovators. The first projects will be high end projects which encourage tourism, such as floating resorts and private floating islands, but eventually the architect plans to benefit the wider society by building affordable floating communities.

Rendering by Waterstudio.NL and Dutch Docklands

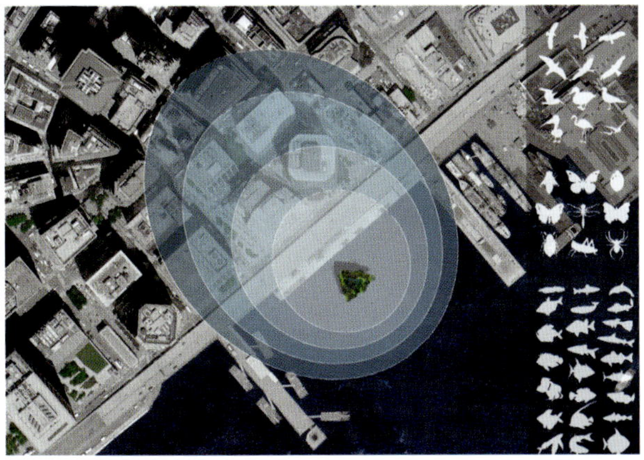

SeaTree

Green is under pressure in high density cities. Instant green floating solutions provide a habitat to fishes, birds, and insects and have a positive effect on the ecology of cities. Our architecture is based on existing oil industry technology.

Rendering by Waterstudio.NL

Ruan Hao
Principal

18/F, South Jing Mao Tower Street
466 Yan An Road
Hangzhou 310006
Zhejiang Province
China

www.lycs-arc.com
hao@lycs-arc.com
+86 (571) 86615392

In terms of architecture, I think the ideal way to deal with the prevailing conditions is to use minimum and restrained design techniques to create maximal and unexpected effects. I call this accidental architecture. This minimum does not mean the minimum in effort—in fact it takes an even longer time than normal and more consideration and design expertise. The way we do this is to create a geometric threshold that is related to its normal condition, meaning that you can always trace back to its original status, but it has to be achieved by restrained geometry, which follows the rules of gravity and the overall structure.

YCS Arc

hitecture

LYCS Interior

This new office design is a renovation of the penthouse floor of a tower in downtown Hangzhou, with a panoramic view of the West Lake UNESCO World Heritage site. Our design juxtaposes the original abandoned storage roofspace and the new office, using a composition of material homogeneity and heterogeneity, and a temporal attenuation of the space. In celebrating the differentiation between an authentic architectural language and wasted space, we retained the imprint of several original elements as a reminder of the leftover space. This project is not a normal vision of minimalism, as it reinvents the notion of historical space via wasted space, taking a critical position on the contemporary tenets of the fast-paced real estate market in China.

Photography by Hao Ruan

CATable

People who live with a cat often speak of taking the cat from their lap like a sentimental ritual of temporary farewell. Equally, they say that an inviting hole can be irresistible to their cat. Their curiosity is greatly satisfied through repetitively exploring the unknown path behind the hole. The design of CATable was a fusion of those experiences, as well as a locus where the interaction occurs. It is a table for us, and a paradise for cats.

The theme of my design is "Accidental Architecture." It is about when a young firm is growing and confronting the architectural conditions in China's current environment. Accidents mean a change to the normal trajectory and imply a turn to common understanding. In China we were facing far more potential accidents than we might have in any other place in the world because we were moving so fast and building so much. It is actually just these type of accidents that help us to improve the system, enhance the architectural environment and try to move forward.

Photographer by Zhenyu Lai

Tian Tai No. 2 Primary School

The project for the Tian Tai No. 2 Primary School strives for a unique design, which will serve as a model school providing a beautiful environment for the cultivation of knowledge, culture, physical fitness, art, and ethics for elementary school children. Our design focuses on the relationship between architecture and site, site and city, and form and function. Because of the very small lot, the 656-foot (200-meter) running track was projected onto the roof level, giving an additional 32,000 square feet (3,000 square meters) of usable area on the ground, as well as the oval shape of the school building, creating a sense of inwardness and security for the students. Lifting the running track also allows for a total project height of four floors instead of five as originally required, creating a more harmonious relationship between the new school and the surrounding urban context. Lastly, we rotated the building by about 15 degrees, creating smaller pockets of space between the site wall and the exterior envelope.

Photography by Shengliang Su

Ma Yansong
Founder and Principal Partner

Dang Qun
Principal Partner

Yosuke Hayano
Principal Partner

7 Ban Qiao Nan Xiang
Bei XinQiao
Beijing 100007
China

www.i-mad.com
press@i-mad.com
+86 (10) 64026632

MAD Architects is currently led by Ma Yansong, Dang Qun, and Yosuke Hayano. Founded by Ma Yansong in 2004, we are a global architecture firm committed to developing futuristic, organic, technologically advanced design, which embody a contemporary interpretation of the Eastern spirit of nature.

With its core design philosophy of Shanshui City – a vision for the city of the future based in the spiritual and emotional needs of residents – MAD endeavors to create a balance between humanity, the city, and the environment.

We have been commissioned by clients of various backgrounds to design mixed-use complexes, museums, theaters, social residences, renovations, artworks, and urban plans. Current ongoing urban projects include Chaoyang Park Plaza, a mixed-use complex project located in the new CBD of Beijing; Nanjing Zendai Himalayas Center, a city-scale urban development with an approximately 6.5 million-square-foot (600,000-square-meter) floor area; and Huangshan Mountain Village, a master planning and architectural design project covering 4.8 million square feet (450,000 square meters). Our signature cultural projects include Harbin Cultural Island (under construction), Pingtan Art Museum (in the design development stage), Ordos Museum (completed in 2011), and China Wood Sculpture Museum (completed in 2011).

MAD has ongoing international projects located, in Rome, Paris, Japan and Beverly Hills, Los Angeles. In the heart of Rome, we are designing a complex multifunctional courtyard building, which contains apartments, office space, and a chapel with a total usable area of approximately 301,400 square feet (28,000 square meters).

MAD Ar

In Paris, our practice was placed first in a competition of 96 teams to build a residential complex on the right bank of the River Seine. In Japan, we delivered an organic design for a local education center. The firm's latest commission is a 140,000-square-foot (13,000-square-meter) mixed-use residential complex in the center of Beverly Hills.

In addition to architecture, we have been a pioneer in contemporary art and design and have participated in significant exhibitions in the 10th, 11th, and 12th Venice Architecture Biennales. Our practice has also participated in exhibitions at the Victoria and Albert Museum (London), the Louisiana Museum of Modern Art (Copenhagen), and MAXXI (Rome). An array of our architectural models have been acquired by the well-known M+ Museum (Hong Kong) as part of their permanent collections.

Our advances have been chronicled in a series of books: *Mad Dinner*, *Bright City*, *Ma Yansong*, and *Shanshui City*. As principals, we have been invited for speaking engagements at the American Institute of Architects, the Architectural Association School of Architecture (London), the World Economic Forum in Davos, the Strelka Institute (Moscow), and the UCL Bartlett School of Architecture (London).

chitects

Dior
GUCCI
DKNY

Nanjing Zendai Himalayas Center

View of vertical city plaza (top); View from lobby in the air (middle); View of sky lake (bottom); View of central valley (left)

Rendering by MAD Architects

Chaoyang Park Plaza

Chaoyang Park Plaza is located in the central business district (CBD) of Beijing, and is composed of over 1.3 million square feet (120,000 square meters), of commercial, office, and residential buildings. The site is on the Southern edge of Chaoyang Park, one of the largest public parks in Beijing. Its proximity to the park will not only create breathtaking views of the city, but will also highly impact the skyline of Beijing.

Like the tall mountain cliffs and river landscapes of China, a pair of asymmetrical towers creates a dramatic skyline in front of the park. Ridges and valleys define the shape of the exterior glass facade, as if the natural forces of erosion wore down the tower into a few thin lines. Flowing down the facade, the lines emphasize the smoothness of the towers and its verticality. The internal ventilation and filtration system of the ridges draw a natural breeze indoors, which not only improves the interior space but also creates an energy efficient system.

The project was awarded the "Leadership in Energy and Environmental Design (LEED)" Gold certificate by U.S. Green Building Council.

Renderings by MAD Architects

Harbin Cultural Island

Exterior

Photography provided by MAD Architects (right); Photography by Xia Zhi (middle); Rendering by MAD Architects (bottom)

Iker Gil
Founder and Director

1564 North Damen Avenue
Suite 204
Chicago
IL 60622
United States

www.mas-studio.com
iker@mas-studio.com
+ (312) 401-4537

Reusing underutilized assets of the city. Exploring the collective. Proposing flexible frameworks to address changing conditions. Instigating the exchange and debate of intellectual positions. Communicating ideas across audiences. While the medium that we use for each project varies (proposals, installations, built work, publications, or exhibitions), the interests mentioned remain a common theme.

With the work of MAS Studio, we are interested in exploring existing but underutilized assets of the city. Those assets range in scale from park systems intersecting several neighbourhoods, to an underground pedestrian system under downtown Chicago, or a single vacant lot. Our aim is to go beyond the distinction of public/private assets, defining new areas for collective interaction. Areas that have the potential to combine public and private assets benefiting from each other, blurring programmatic, ownership, and management boundaries.

MAS

Our proposal for the Chicago Pedway system is a visual exercise to explore new possibilities for the existing underground system. Instead of replicating the qualities and programs of the street level, it proposes a new collective scenario where private and public activities mix, collide, and complement each other. *Cut. Join. Play*, a temporary installation in Little Village (Chicago), reactivates four vacant lots by providing a new artificial landscape, inviting community involvement to what was until then a vacant lot.

More than a final singular solution, it is a flexible framework that can, and has to, be appropriated and activated by the community with the activities they need. Inside Marina City, a project we undertook in collaboration with photographer Andreas E.G. Larsson, celebrates Marina City's qualities 50 years since its opening, not by describing its architectural feats but by focusing on its inhabitants and its success in fostering a diverse community. The photographic output, shared through a traveling exhibition initiated at the Art Institute of Chicago, engages audiences beyond architects in order to have a comprehensive and inclusive conversation to discuss the role and potential of architecture.

With the publication *MAS Context*, a parallel initiative to MAS Studio, we are interested in creating a platform that can build, challenge, and advance the knowledge of our built environment by looking at specific topics. Based around a quarterly publication, *MAS Context* looks at a single topic through a curated collection of authors and contributors from different fields with different perspectives. The release of each issue signifies the beginning of the discussion and not the end of it. Several public events, such as lecture series, exhibitions, and pop-up bookshops organized throughout the year, continue conversations around proposed topics or uncover new ones.

Across the different projects, commissioned or self-initiated, we look at the bigger picture of issues and we distil them into their main ideas and necessary actions. We are proactive about proposing ideas, looking for opportunities that are present in the city and overall, being optimistic about the potential of the city. As architects and citizens, we can't be passive users of our own cities.

Studio

Inside Marina City

While the towers of Marina City present a public and easily recognizable profile to the street, my project focuses on the residents of the building and the highly personal living spaces that make it an energetic and diverse community. Marina City's central location, affordable apartments, and unusually wide range of amenities, including large balconies, bowling alley, theater, and marina, have long attracted creative and engaged residents. Inside Marina City explores the relationship between Bertrand Goldberg's rigorous modular framework for the apartment units and the informal development of these interior spaces by residents throughout the history of the building.

Photography by Andreas E.G. Larsson

Inside
Marina City

Chicago Pedway

The Chicago Pedway is a network of tunnels (with some ground-level concourses and bridges) connecting office buildings, governmental buildings, retail stores, hotels, and train stations throughout the Chicago "Loop." Initiated during the 1930s, the Pedway is, in reality, an incomplete system, which has evolved throughout the decades, reducing its access in some areas and eliminating most of its initial activities. Currently, it lacks a clear and comprehensive identity, recognizable accesses, proper lighting, attractive activities and, in general, any quality that makes it an appealing destination for the residents or tourists of Chicago.

Seeing its current condition, what is the true potential of this system in Chicago? Can it add something new that can improve the quality of life of its inhabitants?

We believe that the Pedway system has incredible potential and can become a new collective space in the city: a pedestrian system that is part of a bigger network of urban spaces, with specific architectural and spatial qualities; and a system that allows for the expansion of the activities currently confined in small spaces, as well as the incorporation of new activities that can enrich the cultural and social scene of Chicago.

Photography by MAS Studio

MAS Context

MAS Context is a quarterly journal created by the design office MAS Studio, which addresses issues that affect the urban context. Each issue delivers a comprehensive view of a single topic through the active participation of people from different fields and different perspectives who, together, instigate the debate. Since its foundation in early 2009, we have released 24 issues, and published 400 articles by 428 contributors from 71 cities around the world. We have organized 40 lectures featuring local and national speakers from multiple disciplines, and four exhibitions featuring the work of architects, graphic designers, and photographers. The combination of different communication techniques (essays, photographs, diagrams, videos, and sound bites) and formats (physical, online, and events) allows *MAS Context* to engage with a wide audience, as well as to efficiently and clearly communicate the complexity of each topic. *MAS Context* combines physical and virtual mediums as well as events, formal or informal, which exploit the potential of a collective gathering and discussion. They provide a new reading to the content of the issue from the online and printed versions, starting or continuing to define and establish forums of critical thinking.

Photography by David Schalliol (top left); MAS Context (bottom left and below)

Mehmet Ziya Cetik
Senior Interior Designer

Redondo Beach
CA 90278
United States

www.hok.com
ziyacetik@gmail.com
+1 (310) 993-6050

Design is an endless and intimate journey. First of all, having the opportunity to imagine and deliver creative, functional spaces and products on a daily basis, as a designer, is a fun and very inspiring process. My tour de force is always striving to understand clients' unique needs while keeping the purity of the original design for creating environmentally friendly, visually striking, and functional designs that are timeless.

This process requires a careful and responsible approach to clients and their individual requirements. Every project is completely different and, as a designer, I truly enjoy the constant challenge of addressing and surpassing the unique expectations of clients from various backgrounds, whether it's for a commercial, residential, or product design project.

Another very important responsibility, as a design professional, is to educate people in the community, as well as younger generations—enhancing and refining their awareness of creative, functional, and sustainable environments and products.

As a young LEED AP designer with over 10 years' experience in the design profession, I have had the opportunity to create a diverse, international portfolio. I also had the privilege and pleasure of managing my own practice, which gave me a comprehensive

HOK

understanding of different disciplines, trends, and needs within the broad world of product and interior design. With this strong background, I am constantly seeking new opportunities to participate in the shaping of our future with a tireless curiosity and passion, without forgetting to have fun, of course! I cannot wait to see what comes next ...

Born in 1980, Turkish designer Mehmet Ziya Cetik graduated from Istanbul Technical University in 2005 with a major in Interior Architecture. The rich culture of Istanbul, which links two continents bringing eastern and western influences together, has been instrumental in the development of Cetik's design vision. With more than a thousand-year-old history serving as the melting pot for "modern" and "traditional" lifestyles, the city provided Cetik with the unique ability to work with and have an understanding of clients from eclectic backgrounds. During his time in Istanbul, he had the chance to work with diverse and prestigious clients with highly recognized firms in Europe. Multiculturalism and the timelessness of the vibrant city manifests itself in Cetik's early designs.

After moving to the US in 2009, he attended UCLA Extension's Interior Design Advance Standing Program, where he was awarded a scholarship for his outstanding design project. He later received his Masters in Interior Architecture from California Polytechnic University in Pomona. During this period, he continued working as a freelance designer and attended various design-related seminars as a keynote lecturer, including *Pacific Design Center West Week 2011*.

At the of age 35, Cetik has completed over 60 projects, including a broad international portfolio with high-end corporate, retail, hospitality, and residential clients, as well as product design projects, many of which were featured in select publications.

Currently, Cetik works with HOK's Los Angeles Studio as a Senior Designer with a creative group of people delivering exceptional out-of-the-box design solutions to their clients. His approach to each project is calculated yet punctuated with emotion, resulting in work that demands to be seen, experienced, and enjoyed.

Cetik lives with his wife Asu in a South Bay coastal city. He enjoys traveling, riding his motorcycle, participating in volunteer activities and breeding exotic chameleons, which guide his inspiration on a daily basis.

Annenberg Space for Photography

The Annenberg Space for Photography Skylight Studios is a brand new 5,100-square-foot (474-square-meter) multimedia studio and retail gallery located directly across the park from The Annenberg Space for Photography. Skylight Studios provides an enhanced lecture-viewing atmosphere and experience through state-of-the-art A/V equipment, a concert-quality audio system and HD broadcast capabilities. Skylight can also accommodate photographer workshops and special events.

The long-running *Iris Nights* lecture series, has relocated to its new home: Skylight Studios. The series brings to life the most current exhibit with presentations by exhibit-featured photographers and other notable experts and guest artists. In addition to hosting the *Iris Nights* lectures, Skylight Studios offers visitors an expanded retail gallery and the opportunity to watch original documentary films from past *Photography Space* exhibits. The new Meet & Greet series takes place at Skylight Studios. This brand new program will allow visitors to interact with photographers, photo editors, and other guests.

Unlike the *Iris Nights* lectures, the format for each Meet & Greet event will vary by photographer but no matter the format, attendees will have the opportunity to interact one-on-one with the featured guest. The design started with functionally planning a studio lecture space, adjoining retail space, and the support spaces

serving both of these public areas in an extremely tight awkwardly shaped footprint. The retail space was placed immediately inside the entry and next to the windows that would offer display space and natural light. A shallow sweeping curve accentuates the stage and allows the audience seating to follow that curve providing good sight lines to the speaker and screens on either side. The design was driven by appealing to our client's audience who are photography aficionados and predominantly middle-aged, hip women taking in a cultural event with friends or bringing their children to engage them in the exciting world of photography.

Therefore the aesthetic needed to be modern, progressive, and edgy. The color palette needed to be monochromatic in the whites, grays, and black to allow the photography to be the focal point and color in the space. The name of the studio is based on a special photography lens, so the cash-wrap was designed as a lens and then the curvilinear lines were continued throughout the design of the ceiling and floor.

There is also a special ritual at the end of each photographer interview in which the photographer is asked to sign one of the lenses, so we designed a back-lit wall in which each of the signed lenses is displayed with the name of the photographer engraved on a plaque below.

Photography by Eric Liagnel

THE POWER OF
PHOTOGRAPHY

Angeles Investment Advisors

Angeles Investment Advisors wanted a radical departure from the traditional corporate design of their previous space and had a goal of branding the new space in a bold but professional manner. The design centers around the concept of creating a "financial loft" in which the extent of the tall volume could be showcased while exposing much of the ductwork and sprinkler system.

Features were added that both add visual interest and reflect the nature of Angeles' work style. In place of a receptionist's desk, a sculptural huddle room/visitors' office defined the boundary of the reception area and was clad in glass and a faux concrete plaster finish. The wing logo was interpreted in two graphic wall features: a literal, large-scale version in silver vinyl (located at the entrance to the brokers' work space) and an abstract "X-ray" version of a wing by the designer was installed over a large credenza outside the main conference room.

Bright accent colors of green blue and red enliven the industrial white, black, and gray palette; chosen by the firm's three partners to help define their own specific spaces.

Photography by Eric Liagnel

Erla Dögg Ingjaldsdóttir
Principal Designer

Tryggvi Thorsteinsson
Principal Designer

2324 Michigan Avenue
Santa Monica
CA 90404
United States

www.minarc.com
erla@minarc.com
+1 (310) 998-8899

Minarc is an award-winning design studio recognized internationally for its modern, innovative, and sustainable design solutions. Inspired by the pure, austere beauty of their native Iceland, principals Erla Dögg Ingjaldsdóttir and Tryggvi Thorsteinsson were drawn to Santa Monica, California and the wide open spaces and rich light that sustain their vision—one that blurs the line between indoors and outdoors through an abundant use of natural light, creation of exterior living spaces, and incorporation of architecturally captured views of the natural environment.

Moreover, southern Californians were early to embrace their innovative design ideas and building techniques, as well as the use of sustainable and recycled materials which have become hallmarks of the Minarc philosophy.

From our multi-award-winning Ion Luxury Adventure Hotel, Silfra Restaurant, and Northern Lights Lounge in Iceland, to the Rockenwagner Soufflé Bakery, each of our projects feature simplicity in design, executed in an environmentally conscious manner, with smart material choices which include reclaimed lumber, recycled rubber and glass, and, of all things, reused artificial turf, while refraining from the use of paint and carpeting.

Each of our projects reveals our passion for our planet, a concern for its future, and the joy we take in creating eco-conscious spaces in which people can live, work, and play.

Mina

rc

Ion Luxury Adventure Hotel

In a land of myth and legend, on an island where fire meets ice, surrounded by hot springs, glaciers, and the mystical glow of the Northern Lights, comes a new hotel where nothing comes between you and this otherworldly landscape; Iceland's newest luxury hotel.

Nestled into the slopes of Mt. Hengill, near Thingvellir national park, the hotel sits perched atop a series of high-seat pillars, a homage to Iceland's first settler, Ingolfur Arnarson, who en-route to Iceland in the year 874 threw his high-seat pillars overboard and sailed to where the Gods brought them ashore.

Like an enormous basalt post pile, with its black, lava-like exterior emerging from the mountainside, the new wing appears suspended, as if floating on the cloud of geothermal steam that surrounds it. The Ion, with its use of natural light, outdoor living spaces and panoramic views of the environs, creates a unique sense of oneness, where the boundaries of walls seem to part and beckon nature inside.

Inspired by the environmental surroundings, it incorporates innovative materials, sustainable practices, and the natural features of Iceland into its design. It is surrounded by hot springs, which provide geothermal hot water and energy. The prefabricated, panelized building system exceeds environmentally safe building standards. Driftwood and other natural materials are used extensively in furniture throughout the hotel. Beds and chairs are built from recycled materials. Mirrors are framed with indigenous Birchwood. Lights made of lava and found-wood illuminate the rooms.

Images of the Icelandic horse, the Seljalandsfoss waterfall, and other Icelandic references adorn the hotel walls. Each guest room is detailed with restrained lighting effects and subtle color palettes, as well as an abundance of natural daylight. Large picture windows in each room offer breathtaking views of Lake Thingvallavatn and the surrounding mountains, over which nature's own theatre unfolds as the most spectacular light show, the Northern Lights, comes to life.

Photography by Art Gray

Appleton Living

Appleton Living is oriented so that all of the rooms can enjoy the outdoor living area, which includes a pool, outdoor dining, barbecue, and play court. We designed its orientation in order to maximize passive solar design and natural ventilation.

With a solar chimney to allow hot air to escape during the summer and to heat cold air during the winter, it eliminated the need for mechanical air handling. The design elements are oriented to take optimum advantage of natural light and cross ventilation, to cut down electrical costs. Interior/exterior courtyards allow for natural ventilation, as do the master sliding window and living room sliders. Solar thermal radiant floor heating exists throughout the house, while a heated patio and fireplace for outdoor dining maximize indoor/outdoor living. The entry living room has glass on both sides to further connect the indoors and outdoors.

We connected floor and ceiling materials in an unobtrusive and whimsical manner to increase floor plan flow and space.

Magnetic chalkboard sliders in the play area and paperboard sliders in the kids' rooms transform the house itself into a medium for children's artistic expression. The simple eco-conscious design is focused on functionality and creating a healthy, breathing family environment. We made a conscious effort in using materials only in their most organic form. Material contrasts (such as stone, steal, and wood) make this modern home warm and family friendly.

Photography by Art Gray

Big Little Rock

Big Little Rock is a modest house made to blend into the beautiful natural environment of Iceland—designed for an easy lifestyle. It can be a vacation home or somewhere to live all year round.

Our simple eco-conscious design focuses on functionality and creates a healthy living environment for the family. The home integrates with the landscape and blends with the environment's "body." A conscious effort was made to use materials in their most organic form. It embraces history with forms found in traditional mud houses, and the sod roof and walls form the solid building.

The project reduces its impact on the natural environment by using natural materials in their most organic form. Concrete, wood, and glass unify to create the building. Large and expansive windows and openings are used in the living area to provide abundant natural light, and frame the beautiful landscape and enable a seamless connection with nature. Sustainable practices—such as using recycled materials like rubber tyres—are used on the kitchen wall.

The home is open and glass is used to make nature fill the house. The summers are filled with endless sun while the winters glow with the Northern Lights and the snow is so bright it lights up the endless darkness.

Photography by Art Gray

Lyndon Neri
Founding Partner

Rossana Hu
Founding Partner

88 Yuqing Road
Xuhui District
Shanghai 200030
China

www.neriandhu.com
press@neriandhu.com
+86 (21) 60823777

Neri&hu (Lyndon Neri and Rossana Hu) strongly believe, as an architectural duo, in architecture as a paradigm for cultural production and social change, aiming to create meaning and significance through their highly contextual and inter-disciplinary approach to design.

"We do not ask to be eternal beings, we only ask that things do not lose all their meaning." – Antoine de Saint-Exupéry

We established our practice in Shanghai, as colorful and confusing a place as one could hope for. The massive rapidity of Shanghai's urban construction forces us to confront a city disconnected from its memory, a city that prefers deeds over deliberation. This somewhat unfortunate reality turns into an opportunity for us, provoking new ways of thinking about the issues of adaptive re-use and historic restoration, which challenge traditional modes and ideologies. We believe that urban fabric and architectural memory should be preserved with a critical and sensitive approach that exemplifies a zeitgeist within the specificity of our context, and is not merely reproduced. Through a delicate handling and negotiation of contradictions, we have shown in many built works the possibility of creating unexpected spatial experiences in historic buildings, giving them new life.

neri&hu d

and resea

We recognize the importance of context, and we are also keenly aware that architecture has never arisen from purely rational thinking or directly as a result of its context. It has always reflected personal and cultural aspirations and ideals, and we do not shy away from that heritage. Our constructed world enables us to understand and remember who we are, and embedded in that identity is not only our shared Chinese heritage, but also our education, mentors, continual exposure to ongoing discourses, and personal obsessions. We have indeed been obsessed with our cultural identity and its representative value both within the national borders and on the global stage. We operate on many shores with the desire to challenge China's misrepresented image, while contributing to the definition of a re-interpreted Chinese aesthetic.

As architects we believe that working under such complex conditions necessitates a broader definition of our scope as designers. We believe in diversity and its profound contribution to creating an inter-disciplinary methodology for design. Because new sets of contemporary problems relating to buildings now extend beyond traditional boundaries, our practice embraces architecture's supremacy in design thinking on all levels. Through our 10 years of practice, we have exhibited extensive work in the realms of product design, graphic design, interior design, alongside our architectural explorations. All the while, the architect's perspective on space and objecthood is always at the center of our investigations.

We insist on research as a design approach as each project bears its unique set of related issues. A critical probing into the specificities of program, site, function, and history is essential to the creation of rigorous design work. Based on research, we desire to anchor our work on the dynamic interaction of movement, detail, materiality, form, and light rather than adhering to a formulaic style. The ultimate significance behind each project comes from how the built form creates meaning through its physical and metaphysical representations.

esign

rch office

Rethinking the Split House

The magical lane houses, which were once the dominant fabric that made urban Shanghai the intoxicating place that it was in the 1930s, are now gradually being demolished, taken over by high density developments all over the city. Our strategy as architects was to rethink the typology of the lane house, keeping the split level formation —a typical trait to lane houses in this city—and add spatial interest through new insertions and skylights to accentuate the architectural integrity of such a typology, contemporizing it for today's lifestyle.

Photography by Pedro Pegenaute

The Waterhouse at South Bund

Located by the new Cool Docks development in the South Bund District of Shanghai, the Waterhouse is a four-story, 19-room boutique hotel built into an existing three-story Japanese Army headquarters building from the 1930s. The boutique hotel fronts the Huangpu River and looks out across the gleaming Pudong skyline. The architectural concept behind our renovation rests on a clear contrast of the old and new.

Photography by Pedro Pegenaute

Design Republic Design Commune

Design Republic Design Commune, located in the center of Shanghai, envisions itself as a design hub, a gathering space for designers and design patrons alike to admire, ponder, exchange, learn, and consume. It houses the new flagship store for Design Republic, a modern furniture retailer, alongside a mixture of design-focused retail concepts, including books, fashion, lighting, accessories, and flowers. The Commune will also have a design gallery, an event space, a café, a restaurant by Michelin-Starred Chef Jason Atherton, and a one-bedroom Design Republic apartment.

Photography by Pedro Pegenaute

Kunlé Adeyemi
Founder

Lagos, Nigeria, Amsterdam,
The Netherlands

www.nleworks.com
contact@nleworks.com

Kunlé Adeyemi is an architect, urbanist and designer. His recent work includes "Makoko Floating School," an innovative prototype floating structure located on the lagoon heart of Nigeria's largest city, Lagos. This acclaimed project is part of an extensive research project *African Water Cities* being developed by NLÉ, an architecture, design, and urbanism practice, founded by Adeyemi in 2010 with a focus on developing cities. Currently, NLÉ is developing a number of urban, research and architectural projects in Africa; one of which is Chicoco Radio Media Center; the amphibious building in the Delta city of Port Harcourt in Nigeria.

Born and raised in Nigeria, Adeyemi studied architecture at the University of Lagos where he began his early practice, before joining Office for Metropolitan Architecture (OMA) in 2002. At OMA he led the design, development, and execution of several large and prestigious projects in Europe, Asia, Africa, and the Middle East. These include the Shenzhen Stock Exchange Tower in China, the Qatar National Library in Doha, and Prada Transformer in Seoul. Adeyemi is currently one of five members of the International Advisory Council for the World Design Capital 2014, a Fellow of the World Technology Network, and an international speaker, lecturer, and the Baird Visiting Critic at Cornell University, Ithaca, New York.

One of the "mega century's" dominant and unstoppable trends is urbanization. The outcome is a growing number of megacities worldwide, all of which face the same challenges. Within this

NLÉV

context, just as Silicon Valley acts as the home for new technologies, it is the cities of the developing world that will generate responsible solutions for the larger world. As thinkers, creatives, and agents of change, NLÉ's role is to reveal these solutions and apply them so that we shape the physical, human, and commercial structures that are critical to the near future of human civilization.

Our activities encompass city development research and strategy advisory services, conceptualization and creative structuring, architecture and product design, infrastructure design, and arts and cultural urban interventions. Across all these fields we explore the infusion of relevant global ideas and the advanced technologies that add sustainable value. And our global and diverse network of experts and collaborators, as well as our products and our services, are of world-class quality and integrity. At the same time, we ensure authentic relevance through the deployment of local resources, capacity building, and economy of means. In this way, we act in a new language for shaping and advancing the multiple physical and human architectures of the "mega century."

NLÉ means "at home" in Yoruba, the language of Africa's first truly urbanized population. From the 11th century onwards, the Yoruba lived in a network of West African cities characterized by sophisticated commercial and governing structures. But within NLÉ's philosophy, the home is much more than walls, floors, and ceilings. Instead, it refers to the fundamental building blocks of the city, to everyday life and the uses of public space in the emerging and endlessly complex urbanisms of developing regions.

Due to rapid urbanization and the forces of globalization, the people in such cities have redefined our understanding of cities through their innovative and economical appropriation of spaces, materials, and infrastructure. The environment's streets, bridges, sidewalks, raw materials and junk are acquired, cultivated, and reconstituted to achieve maximum necessities through minimum means. This creates unique organizations and advanced social structures in which people live, work, trade, and play in ways that are now rarely seen in the "developed worlds," which are now ironically the "less developing world."

We believe that these colorful and seemingly chaotic worlds hold some of the solutions to the problems faced by cities in the century of "mega needs." Our aim is to work with, and be inspired by, these worlds to create more social, political, and economically responsible interventions for and with the people who have made them home.

Portrait Photography: Reze Bonna

Vorks

Makoko Floating School

Makoko Floating School is a prototype structure that addresses physical and social needs in view of the growing challenges of climate change in an urbanizing African context. It is a movable "building" or "watercraft" currently located in the aquatic community of Makoko in the lagoon heart of Africa's most populous city—Lagos, Nigeria. It is a floating structure that adapts to the tidal changes and varying water levels, protecting it from flooding and storm-water surges. It is designed to use renewable energy, to recycle organic waste, and to harvest rainwater. Like many homes in Makoko, the primary school building is structurally precarious and susceptible to recurrent flooding. Sadly, the inability of the building to effectively withstand the impact of increased rainfall and flooding has frequently threatened local children's access to their basic need—the opportunity of education.

In response to this and in close collaboration with the Makoko community, NLÉ has developed a prototype floating structure that will serve primarily as a school, whilst being scalable and adaptable for other uses, such as a community hub, health clinic, market, entertainment center or housing. The prototype's versatile structure is a safe and economical floating triangular frame that allows flexibility for customization and completion based on specific needs and capacities.

Photography by NLÉ

Chicoco Radio

Chicoco Radio is a floating media platform being built with and for the residents of Port Harcourt's waterfront community in Nigeria. The structure is conceived as a linear composition of public spaces from land to water: a community radio station, recording studios, computer center, meeting rooms, amphitheater and cinema. The radio broadcast mast is an integrated architectural component raising the structure like a bridge: launching one end of the building into the water, suspending the other in the air. The waterside of the building is a floating stage and jetty responding to the ebb and flow of the tide. The airside is the broadcast space where programs and music are made to air. The cantilevered studios open a shaded landscaped area beneath them—open public space beneath a place of open public debate.

Through a deeply responsive and collaborative design process, local residents have provided valuable insights to this solution, which carefully addresses their challenges and strongly reflects their collective aspirations. Chicoco Radio will be built, owned, operated, and maintained by the waterfront communities.
As a "bridge to transformation," the amphibious nature of the building offers a reconnection between the communities' life on land today, their historic past and their potential lives on water in the future.

Images by NLÉ

CDL Microfinance Bank

Credit Direct Limited is one of the fastest growing microfinance banks in Nigeria, providing microcredit to low to medium income earners. Situated in the Ikeja district of Lagos in Nigeria, the new headquarters for CDL consolidates its various facilities into a single building that redefines the conventional notion of a bank building as a closed and private institution, by creating a more approachable, open and civic space to meet the needs of its customers—the everyday people.

The specific tropical climate of Lagos suggests an alternative approach to designing a building with distinct private and public functions. A conditioned volume containing the most public functions—efficiently enclosed and protected to reduce heat gain—is suspended above the ground floor comprising an auditorium, entrance lobby, and service block. The top floors with a more private function have a transparent envelope which increases the opportunity for views and natural ventilation. The envelope is protected by a generous folded roof plate that also harvests rainwater.

These distinct spaces are connected in a continuous loop that starts with a lush garden on the ground level, leading up to a roof garden via an open walkway, connecting the various functions and providing direct access to the restaurant café.

Images by NLÉ

Mitchell Joachim
Professor and Co-founder

63 Flushing Ave, Unit 192
Navy Yard, Bldg 280, ST E 814
New York
NY 11205
United States

www.terreform.org
mj@terreform.org
+1 (617) 285-0901

My research projects are structured at the union of wet-lab and computational biology for urban architectural design. In the most fundamental of interpretations this is the study of socio-ecological design. A nascent term that recognizes science based design alone is not enough to confront the enormous issues facing humanity with regards to climate dynamics.

It is my understanding that no matter how comprehensive the solutions are to climate change, the social constraints are a considerable factor. Even if the key factors in both science and design are tenable, the capricious public may still have an opposite perspective. I've been making headway in this self-defined interdisciplinary approach for almost two decades.

I've explored these issues at three primary scopes of inquiry: mobility, architecture, and urbanization. For the last five years, I've been designing environmentally driven transportation systems, buildings, and urban neighborhoods. They are all cross-linked and integrated with one another at different scales of operation to mitigate issues of waste, energy, water, materials, sustenance, and air quality. A few key examples are: the Fab Tree Hab living prefab homes, Mini-stacking electric cars for China, Mycoform structures grown from strains of fungi, a Gen2Seat biopolymer compostable chair, an Urbaneering Governors Hook resilient waterfront infrastructure made of decommissioned military ship hulls, and the Bio City Map of 11 Billion People, using genetically modified E. coli colonies as printed geographies.

NYU/Terre

Most of my research is completed in conjunction with my 501c3 nonprofit studio, Terreform ONE, in the Brooklyn Navy Yards, New York City. It is an architecture and urban think tank which advances socio-ecological design in derelict municipal areas. By formulating unsolicited feasibility studies and egalitarian designs, our mission is to illustrate speculative environmental plans for New York City and other cities worldwide.

My mission is to support community outreach and master plan solutions in underprivileged areas which do not have direct access to qualified architects and urban designers. I direct a pro bono studio and laboratory for adept urban specialists to investigate and refine the broader framework of ecological city design. My organization strives to inspire unique solutions for global sustainability in the form of smart infrastructure in an equitable economy. Through exhibitions, publications, and lectures, the group seeks to encourage public appreciation for the numerous methods in which architecture and urban design improve human life.

formONE

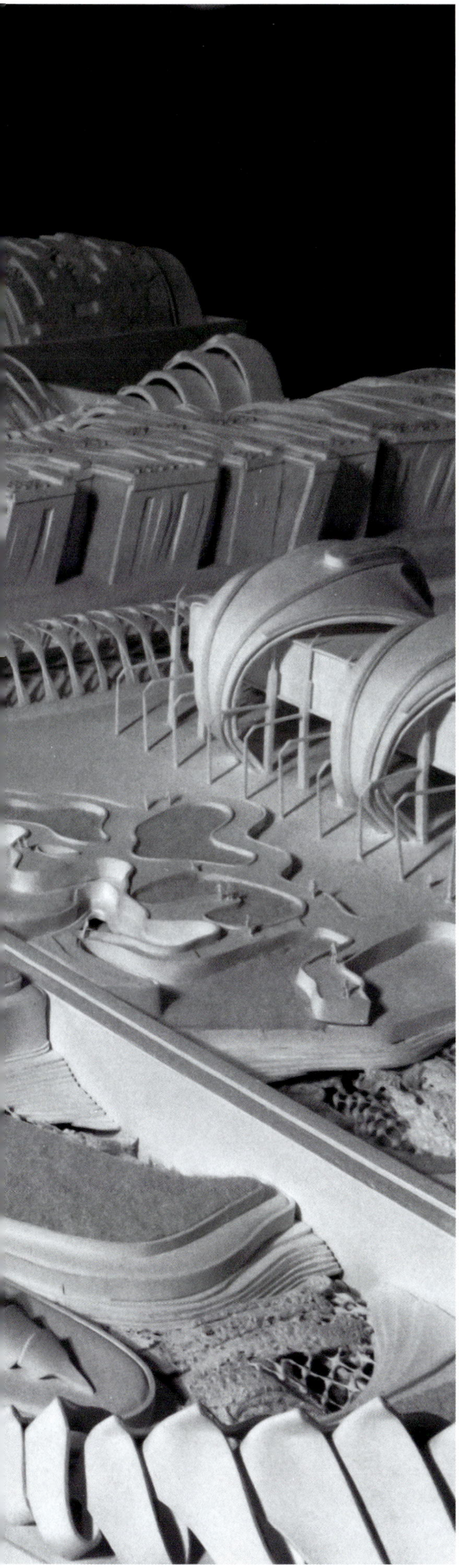

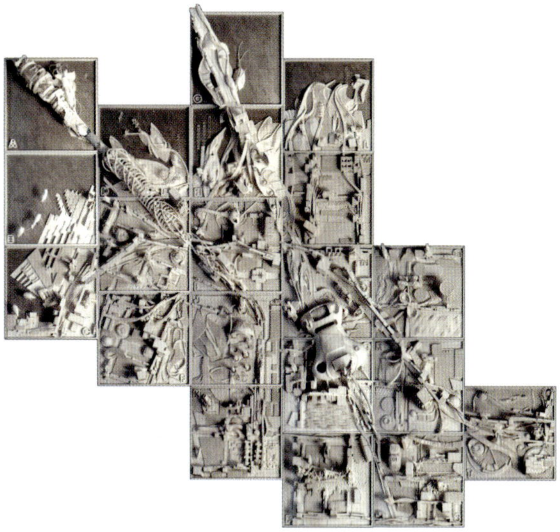

Urbaneering Brooklyn 2110 Self-Sufficient City

Our primary assertion for Brooklyn 2110 is that all necessities are provided inside its accessible physical borders. We have designed an intensified version of Brooklyn which supplies all vital needs for its population. In this city, food, water, air, energy, waste, mobility, and shelter are radically restructured to support life in every form.

The strategy includes the replacement of dilapidated structures with vertical agriculture and housing merged with infrastructure. Former streets become snaking arteries of livable spaces embedded with renewable energy sources, soft cushion based vehicles for mobility, and productive green rooms. The plan uses the former street grid as the foundation for new networks. By re-engineering the obsolete streets, we can install radically robust and ecologically active pathways. These operations are not just about a comprehensive model of tomorrow's €™s city, but an initial platform for discourse. We think the future will necessitate marvelous dwellings coupled with a massive cyclical resource net. The future will happen; how we get there is dependent upon our planned preparation and egalitarian feedback.

Photography and images by Mitchell Joachim, Terreform ONE

Fab Tree Hab: An Edible Prefab Home for Humanity

This home concept is intended to replace the outdated design solutions at Habitat for Humanity. We propose a method to grow homes from native trees. A living structure is grafted into shape with prefabricated Computer Numeric Controlled (CNC) reusable scaffolds. Therefore, we enable dwellings to be fully integrated into an ecological community. 1. Composed with 100% living nutrients. 2. Make effective contributions to the ecosystem. 3. Accountable removal of human impacts. 4. Involve arboreal farming and production. 5. Subsume technology into terrestrial environs. 6. Circulate water and metabolic flows symbiotically. 7. Consider the life cycle, from use to disposal.

Photography and images by Mitchell Joachim, Terreform ONE

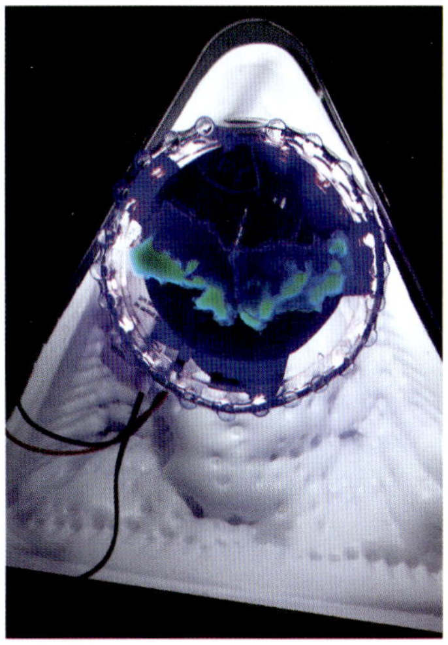

Bio City Map of 11 Billion

In the next 100 years we can expect human population to reach 11 billion people. What does this increased massive growth look like? We used a Dymaxion map grid to communicate an all-encompassing view of world population density in cities by means of data. The map visualizes the earth as one entire urbanized place, instead of unconnected settlements, towns, municipalities, and disparate regions.

Our Bio City Map displays population density as a parametric graph on the front, while the back is made with living biosynthetic matter. These living elements focus on numerous mega-city inhabitants, genetically designed and grown inside petri dishes. Our novel approach experimented with living populations which consisted of hundreds of thousands of bacteria colonies. We preferred to show population density in graph form with actual colonies that were alive to challenge typical computer driven processes.

We chose colonies of E. coli as a method of analog computation using synthetic biology. Population density was represented in two different forms of bioluminescent E. coli under UV light. Glowing red E. coli represented future census projections, while green represented existing demographic conditions that you would find in cities. We used the dilution method in biology to show the range of densities of E. coli populations in each petri dish. Micro-stencils derived from CAD files shaped the E. coli into specific geometries which display the current geopolitical boundaries in cities. This is an interdisciplinary project which involved cartographers, urban planners, biologists, and architects, which completed a manifestation of the near future for human population density. We argue that most nations cannot view the effects of planetary population density through the lens of just one city or region. Instead we aimed to reveal the long-range effects of immense human population in areas of present and speculative urban intensity.

Moreover, we expanded the technique of "bacteriography" (bacteria photography) to shift scale and underscore the highest zones of growth.

Ultimately, the bacterial shapes grow to reveal variant patterns of biological transformation in urban regions. By using biosynthetic based materials, we expect to narrow the gap between idealized mathematical interpretations and observable events in nature.

Our team consisted of a consortium of individuals trained and/or working at the Harvard University Medical School, Harvard Graduate School of Design, MIT Media Lab, New York University, Cooper Union, and local non-profit organizations.

Photography by Mitchell Joachim, Terreform ONE

Kersten Geers
Partner

David Van Severen
Partner

251 Avenue Louise
B–1050 Brussels
Belgium

www.officekgdvs.com
mail@officekgdvs.com
+32 2 503 05 71

OFFICE Kersten Geers David Van Severen is renowned for its idiosyncratic architecture, in which utopian and unrealized projects are also customary. We do not invent architecture, but reflect and consider what architecture can signify and be today. We aim to reduce architecture to its bareness and essence. Our architectonic ideas start from geometrical corrections and rather rigid classifications, in order to measure the world as it presents itself to us and allow life to unfold in all its complexities.

"All of my art has been premeditated: having a notion of the end and not the means to the end. The means to the end has always been secondary in my art. It's the end product that I'm after." –Ed Ruscha

OFFICE Kersten Ge
David Van S

OFFICE Kersten Geers David Van Severen was set up in 2002 by Kersten Geers (born in 1975) and David Van Severen (born in 1978).

Kersten Geers graduated in Architecture and Urbanism at the University of Ghent, Belgium and at the Esquela Tecnica Superior de Arquitectura in Madrid, Spain. He worked with Maxwan Architects and Urbanists and with Neutelings Riedijk Architects in Rotterdam. He was professor at the University of Ghent, and visiting professor at Columbia University, NYC, and the Academy of Architecture in Mendrisio (CH), and is currently teaching at the EPFL, Lausanne (CH). He is a founding member of the architecture magazine San Rocco.

David Van Severen graduated in Architecture and Urbanism at the University of Ghent, Belgium and at the Esquela Tecnica Superior de Arquitectura in Madrid, Spain. He worked with Stéphane Beel Architects, Xaveer De Geyter Architects and for Atelier Maarten Van Severen. He was a teacher and guest critic at the Academy of Amsterdam, the TU Delft, and the Berlage Institute in Rotterdam, and he's currently guest tutor at Architecture School of Versailles.

ers
Severen

Office 62: City Villa

In order to double the surface of a seemingly freestanding house in this strangely lush environment, we decided to underline the existence of everything already there. We projected the new addition under the existing house, not adding any new volume, but effectively creating its pedestal. The pedestal turns the existing house into an exhibited object: undone of its original importance.

The existing house becomes a night house, a ghost house on top of a new, excavated villa.

It is a house that we designed as a set of different spaces traced by columns. The column rhythm defines plan and sequence: a set of spaces which are not functionally defined. We made the structure from massive concrete beams and columns, measured with maximal tectonic effect.

The structure is translated into a spatial idea. Concrete columns become stained wooden columns as soon as one crosses from inside to outside, effectively creating a spatial sequence of hypothetical places to stay. The villa thus exists in the green island that is maintained by its very existence, a conscious contribution to an urban tissue on the verge of extinction.

Photography by Bas Princen

Office 39: Villa Buggehout

This freestanding dwelling is situated in the middle of a parcel of lands between woods and an agricultural area in Belgium. An enclosing modular steel fence is an integral part of the design and defines the volume of the house. The surrounding unkempt garden functions as automotive access around the house. The dwelling itself is composed of two levels: an open "outside house" on the ground floor and a closed "inside house" on the first story. The outside house is conceived as a patio villa with a garden, its thick double walls carry a concrete platform forming the base of the inside house.

The inside house is a compact set of rooms in the depth of the roof of the patio villa. All detailing on the project is designed from the inside out: the huge sliding windows are added to the façade in such a way that the frames are invisible, with views to the still, open, rural landscape, while the neighbouring houses remain relatively invisible.

By deliberately moving the fence in from the edges of the property line, the dwelling becomes "freestanding"—a rare luxury in crowded little Belgium.

Photography by Bas Princen

Office 56: Weekend House

This weekend house occupies the complete area of a deep lot in a small town in Flanders. We transformed the long backyard into the weekend house proper; the whole length becomes the house, turning the existing building into a guest house. The new house is organized as a perspectival sequence of four equal rooms, thus providing a sequence of similar spaces whose infill is always different. Each of the rooms gets its specificity through its particular furniture. Together they form a set of interlocking mini-universes.

The core of the house is formed by the middle two rooms: one with a fixed roof, where two wooden elements make the space inhabitable; and the other is a tropical greenhouse with a mobile roof, which contains a bathroom unit and a swimming pool. On each side of these two rooms one finds a garden room. Closest to the living room is a simple green landscape, and between the green house and the guest house a paved courtyard. In the summer the mobile roof moves over the court, turning the sequence of the four rooms into a complete house, effectively creating a weekend house cut away from any sense of context and reality: a mirage, in effect.

Photography by Bas Princen

Li Hu
Founding Partner

Huang Wenjing
Founding Partner

Building A103,
No. 46 Fangjia Hutong,
Dongcheng District,
Beijing 100007,
China

www.openarch.com
office@openarch.com
+86 (10) 84407533

Architecture is not just about aesthetics, image, light, materiality, details, spirituality, social responsibility, sustainability, theory or history... It is about all of them, but most important of all to us, is its innovative power to transform people and the way they live, while striking a new balance between the man made and nature.

We live in such an era, that Chinese urban history is quickly being rewritten; our future is being built in haste, often without either careful planning or conscientious execution. As architects who choose to practice in these unprecedented social and economic contexts, where ample opportunities go along with great difficulties, we actively seek design strategies that tackle the problems and respond directly to the challenges of our time. Public and Nature are two underlying key words in our practice.

Imperial Chinese cities are well-planned walled-in cities with rigorous order and beautiful forms. Yet the order is to serve one purpose: the need of absolute control by the privileged class. Throughout Chinese urban history, public spaces have been almost non-existent. In our age of globalization and China's rapid journey into modernity, the historical urban structure is no longer functional; but a new structure is yet to be determined. We are in urgent need of cities that are open and adaptable to modernizing Chinese society, where lives confined in courtyard walls are no longer sufficient. We see the great needs of public spaces in all kinds of forms and scale, all open and free, serving society at large.

OPEN Arc

In our urban projects, we strive to create spirited public spaces with innovative strategies. Whether lifting the building off the ground to provide gardens freely accessible to all, or injecting public activity spaces into facilities that would be otherwise monofunctional, Innovation is the key. Through profound observation and research, we set out to create exciting new spatial forms, which can creatively reorganize people's rapidly modernizing social lives while accommodating Chinese cultural and social needs.

Another great challenge facing us is the rapidly deteriorating ecological environment. Green is no longer just a fancy concept. It is the cornerstone of our practice. We build with reverence and sincere respect to nature. We try our best to reduce the disruption to any given site, while taking full advantage of the specific conditions. Working with lead consultants in the field, innovative solutions for both passive and active energy saving strategies are integrated into our designs.

We see it as our mission to create poetic architectural and urban spaces, which encourage healthy social exchange; and our responsibility to find the ultimate balance between nature and people through building practice. In other words, to build leaving a minimal footprint and maximum social impact. Architecture is our vehicle. Through innovation we see the power of architecture and its potential to effect the greatest social and environmental transformations ever seen.

hitecture

Beijing No. 4 High School Fangshan Campus

The design of the school began with deep observation and reflection on the current state of Chinese education and urban conditions. Through a series of innovative and practical design strategies, we created a joyful and effective learning environment between a rich landscape and multi-faceted social spaces.

Confronted with an energy crisis, global warming and frequent natural disasters, we feel that it is an urgent issue to teach our younger generation how to respect and protect nature. In the design of the new school, conventional teaching spaces such as classrooms and laboratories are organized within a rhizome-like structure, which provides eco-friendly and efficiently connected spaces. The roof of the teaching building is utilized as an organic farm for the students to learn the basics of farming. Those special spaces, such as the auditorium, gym and canteen are located partly underground. The undulating roof of these large volumes become part of the ground, free from buildings above, forming continuous gardens and activity spaces for the new campus. Thus in this new school, teaching and learning are uniquely set inbetween the farm on the roof and gardens on the ground.

We think social interaction is especially important in adolescents growing up in the internet age. In the new school, a large number of public spaces of varying forms and scales, as well as different levels of privacy, are integrated into the design, encouraging the inhabitants to wander, play and meet. Classrooms are extended to the outdoors, while playgrounds are brought into the interior. While increasing the efficiency of space usage, they make learning a dynamic experience full of fun and discoveries. It is our hope that the open campus will inspire positive changes for the Chinese perception and practice of education.

As the first high school built according to green building 3-star standards in China, our design takes full advantage of passive energy saving solutions such as exterior sunshade, a green roof and natural ventilation and lighting. It also employed advanced green strategies to further enhance the sustainability aspects of the project, such as geothermal heating and cooling, solar heating, heat recovery and rainwater recycling.

The clarity of structural expression, simplicity of natural materials and green features all make this new campus building complex a vivid architectural textbook for students.

Photography by Su Shengliang (far left); Xia Zhi (left and below)

OPEN Prototype for Temporary Sales Pavilion

Reflecting on the wasted resources and design efforts behind the sales pavilion as a uniquely Chinese phenomenon, we conceived a prototype Version 1.0 of a standard temporary sales pavilion, which enables speedy construction, is adaptable to different programs and sites, and more importantly, can be disassembled after each use and reassembled at another location.

This re-useable building is an experiment to test the ultimate potential of sustainability, not only in its realized physical form, but also the meaning of its life cycle.

Devised from one modular system, the steel structural skeleton, and the composite floor panels made from steel and concrete, are prefabricated and bolted together on site. Both the solid aluminum panels and the glass panels take on the same module, which allows a somewhat free arrangement of openings on the façade, according to different program requirements behind. Internally, the space is organized with one bridge which connects five boxes that each contain different types of functions. The interior décor can vary, as well as the positions of the five boxes.

This temporary sales pavilion offers an opportunity for us to realize for the first time our concept and experimentation on mass customization, which began almost 10 years ago in New York City in the formative years of our firm.

Photography by Su Shengliang

Gehua Youth and Cultural Center

This is a youth camp experience center and cultural activity center for locals, with open and flexible event spaces adaptable to the ever-changing programs, making it a kaleidoscope-like mini-world. Facing the gently sloped central courtyard, 20-foot-tall (6-meter-tall) folding doors slide open, allowing a small new media theater to instantly transform itself into a large performing space. Boundaries between inside and outside are blurred. A building fused with the surrounding natural environment, combining passive energy strategies with natural materials, it provides a place for youngsters to discover and create their own experiences and stories.

Although the theater is relatively small with about 120 seats, we designed and equipped it to host high quality professional performances. When both sets of folding doors behind the stage are fully opened, the courtyard suddenly becomes a part of the theater, creating a delightful surprise that transforms the small theater into a large performing arena.

The performers act and dance inside and outside, taking in natural elements from the external environment as part of the performance. The spectators enjoy a completely different kind of theater experience, both novel and unexpected. On other occasions, the courtyard that rises along the slope is an ideal auditorium for both performances and open air cinema.

Photography by Xia Zhi

Marcelo Spina
Principal

Georgina Huljich
Principal

3249 Casitas Ave
Suite 214
Los Angeles
CA 90039
United States

www.P-A-T-T-E-R-N-S.net
info@p-a-t-t-e-r-n-s.net
+ (1) 323 284 8816

P-A-T-T-E-R-N-S is an architectural design practice based in Los Angeles with 15 years of building experience. Founded in 1999 and directed by Argentine born Architects Marcelo Spina and Georgina Huljich, P-A-T-T-E-R-N-S gained international recognition for its subtle approach to architecture; one that seamlessly integrates advanced technology within an extensive consideration of form, novel tectonics and innovative materials.

With a decidedly global influence and working across multiple scales, programs and cultures, the office completed projects in the US, South America and Asia. Its work was exhibited and published worldwide most notably at the Venice Biennale in Italy, the Art Institute of Chicago, San Francisco MOMA, Vienna MAK Museum, where their work is also part of the permanent collections.

P-A-T-T-E-R-N-S received numerous prizes and awards including the American Architecture Award, an ACSA Faculty Award, two AIA LA Honor Awards, an AIA National Design Review Honorific Mention; first prizes in the competitions for the SCI-Arc Graduation Pavilion, a Temporary Pavilion for MOCA, the Vertical Garden at the Schindler House, the New SCI_Arc Café and the third prize in the prestigious Young Architect of the Year Award. In 2011, P-A-T-T-E-R-N-S integrated the Emerging Voices Series of the Architectural league of New York, and was the recipient of the "Arch is" Award by the AIA LA Chapter. In 2013, Marcelo Spina and Georgina Huljich were the recipients of the prestigious USA Fellowship.

In addition to their professional and academic work, Spina and Huljich were the curators of *Matters of Sensation* at Artists Space. One of the most important collective exhibitions in architecture that Artists Space ever organized, the influential show opened to great acclaim in 2008.

P-A-T-T-

Current projects include SoLA Village, a 2.5 million-square-foot (232,300-square-meter). Mix-use Project for Downtown Los Angeles and Luxury Hotels for Bequia Island in the Caribe and Rottnest Island in Western Australia.

Marcelo Spina received his BArch from the National University of Rosario, Argentina [1994], and a MArch from Columbia University, New York [1997]. He is a Design Faculty at SCI-Arc since 2001, and the coordinator of ESTm Program. He has been a Visiting Professor at the Universities of Yale, Syracuse, Harvard, Berkeley, Vienna, Innsbruck, and Di Tella among others.

Georgina Huljich received her BArch from the National University of Rosario, Argentina [2001], and a MArch from University of California, Los Angeles [2003]. Huljich is an Adjunct Associate Professor at the Department of Architecture at UCLA and the Director of the Summer Program Institute. She has been a Visiting Professor at the Universities of Yale and Syracuse, a Maybeck Fellow at UC Berkeley and a Visiting Critic at the University of Southern California in Los Angeles.

P-A-T-T-E-R-N-S specializes in architectural design across scales, from small objects to large buildings. As a result, the office produces innovative buildings and projects that are both exclusive and accessible, forward looking and subtle, distinctly contemporary in their formal and material aesthetics and unapologetically timeless in their functional and spatial performance.

Our creative process involves an analytical approach to the specifics of each project, client and context, an imaginative philosophy and unique artistic sensibility to design architectural and urban form and a sophisticated technological solution to its construction and material tectonics. Experience in collaboration with consultants in diverse fields ranging from economic advisors, master planning, to structural engineering and material experts allows us to produce increasingly unique projects locally and abroad.

By deploying both pragmatism and imagination during all steps of design, the singularity of our architectural approach is that we can fluidly integrate speculative thinking with an effective production. Our vast experience in designing and building architecture worldwide enable us to create projects that maintain a refined balance between theoretical and aesthetic concerns, contemporary demands and long lasting values such as functionality and organization. By closely working with clients, we ensure that their vision is materialized in the most progressive and realistic way, so our final design is both distinctive and sustainable, providing worldwide visibility and recognition.

E-R-N-S

Jujuy Redux

Jujuy Redux is a mid-rise residential building in Argentina, which proposes a formally subtle and spatially complex mass. This operates simultaneously at the scale of the entire building volume, as well as at that of each apartment, challenging over-assumed notions of fixed scalar transformations within mid-rise housing typologies: playing with the envelope being either detached from the units, or connected with them.

Since Neoclassicism, the local code has functionally evolved to prevent any volumetric or ornamental projection beyond the construction line. Turning parapets into slabs and lintels into eaves, our answer to this challenge is to express and complicate the readability of horizontal strata producing effects of diagonal torsion. The repetition of local horizontal articulations produce a linear transition from mass to volume and from volume to surface, wherein a maximum visual and physical distortion is generated with the minimum amount of formal means. As a result, the building dematerializes towards its corner, allowing the living rooms, the social space par excellence of each apartment to fluidly relate to the pedestrian activity in the street below. To be mostly occupied by young couples and students, the project consists of 13 small individual units and a duplex on the terrace, each with two bedroom and two bathrooms.

Photography by Gustavo Frittegotto

League of Shadows

League of Shadows is an outdoor semi-permanent event structure serving the SCI-Arc Campus (Southern California Institute of Architecture) and the Arts District in downtown Los Angeles.

Positioning the pavilion at the corner of 4th and Merrick Street gives the project a double life. Its obvious function as an outdoor public event space and a more iconic one, as a formal beacon at the southern tip of the property—a billboard sign that reasserts SCI-Arc's institutional presence in downtown Los Angeles. The intention of this placement is that, when seen from 4th street, the new pavilion serves as a landmark, a natural marquee for SCI-Arc campus. Due to both its placement and visibility, the pavilion can become a suitable venue for SCI-Arc to engage outside audiences from the downtown community and the broader LA area.

By positioning the project vertically, the pavilion produces enough shadow during the early and late hours of the afternoon (when events are held) to provide comfort zones sheltered from the sun. More importantly, its figural shape acts as a natural envelope for the event, providing students and families controlled views of the event in what it is otherwise a fully open parking lot. Volumetric and graphic, the intent of the project is to assert a certain ambiguity in its formal reading. Starting from the definition of the audience, the required shadow path in the ground is projected upward, producing as one of its possible instantiations, three fingering rectilinear volumes. By depicting simple volumes from the street and a carved unified grouping of enveloping surfaces from the interior, the proposal provides a dual reading: a simple outline from far away and complex radiating superficial texture at close range. The three fingering volumes are carved with three large saddle surfaces. These concave surfaces produce a vaulted interior, which further articulates the organizational sequence of the event. Furthermore, by touching the ground at four points, they envelop the audience producing favorable acoustic and visual conditions for events.

Photography by Monica Nouwens

Chengdu Fluid Core-Yard

Zhixin Hybrid Office Park is part of a large new campus development comprising approximately 20 new buildings, situated on the outskirts of Chengdu in Sichuan Province. The program requested an innovative mixture of private residences, office space, and leisure facilities, allowing a new startup company or a local headquarters to enjoy both proximity and cooperate identity. The project presented the opportunity to speculate in a context with very little context—despite the specificity of each new building, everything is also fairly generic.

The building is organized by a diagonal wedge of circulation that dynamically cuts the plan and connects front and back. The mass is subtracted from at its corners, generating similarly opposed structural cantilevers that produce a strong sensation of levitation. Waffle hyperbolic paraboloids connecting vertical walls and horizontal slabs further induce the sense of obliqueness throughout the building, linking the entry street to a golf course visually and physically. The repetition and inversion of the same hyperbolic geometry at opposing ends produces a spatial reciprocity. Experientially, this geometry creates a sense of déjà vu while rapidly moving in and out of the building, then subtly but substantially subverting the generic mass.

A standard panelized aluminum envelope is used to clad all flat surfaces of the building. The inset eye-like window cavities and the coffered torque surfaces underneath the overhangs are constructed in GFRC. In both cases, their finish is matte plastic with cement coloration. While aluminum planes are panelized, GFRC surfaces are seamless. The interstitial placement of GFRC surfaces and their proximity to the aluminum creates attenuate their contrast, therefore challenging immediate tectonic readings of the building.

Photography by P-A-T-T-E-R-N-S

Patrick Tighe
Principal

1632 Ocean Park Blvd
Santa Monica
CA 90405
United States

www.tighearchitecture.com
patrick@tighearchitecture.com
+1 (310) 450-8823

If anything remains consistent in our work, it is our process—a pushing of extremes until they buckle and reunite into a complementary whole.

We return continually to implied dichotomous zones and "unfix" them.

We draw from them—extending some, subtracting others—then build them up until the essence of each project becomes visible. We work in five areas, each allowing us to move forward with the work in a different way: craft, from the manufactured to the technological; material, from the natural to the synthetic; experience, from the actual to the perceptual; geometry, from the simple to the more complex; method from the analog to the digital.

Our work embraces rather than resolves conflict. We find these conditions to be liberating, and we are not timid about exploring them.

We push dichotomies to find a resolution for building and are interested in how other disciplines and points of view play a role in our architecture.

We insist on the cultural exchange of ideas and are curious about new means of fabrication, materials research and, of course, digital.

We are inspired by theory in drawing and various modes of representation, believing that the culmination of the idea is its physical manifestation, and the greatest reward is the built work.

In every case, we cultivate dichotomy and select the most desirable elements to enhance our projects, toward a more vital architecture.

Patrick Tighe Arc

Patrick Tighe, FAIA is principal and lead designer of his namesake firm, Patrick Tighe Architecture. Committed to creating an authentic, contemporary architecture, the work is informed by technology, sustainability, and building innovation. Since its inception, the firm has produced a strong and diverse body of projects that includes city-developed affordable housing, commercial, mixed-use projects, installations, and award-winning interiors.

The firm has received over 50 design awards, including eight National AIA Honor Awards, American Architecture Awards, a Progressive Architecture Award, Los Angeles Architecture Awards, Westside Prize, Best of Year Awards as well as local AIA Honors. In 2011, Patrick Tighe was elevated into the College of Fellows of the American Institute of Architects, the highest honor bestowed by the AIA. Patrick Tighe was awarded the prestigious Mercedes T. Bass Rome Prize in Architecture, the American Institute of Architects' Young Architect Award and the 40 under 40 Award. Tighe is a Fellow of the American Academy and The MacDowell Colony.

Patrick Tighe received a Master of Architecture from the University of California, Los Angeles. Prior to establishing Patrick Tighe Architecture, he worked in the offices of Frank Gehry and Thom Mayne. Tighe is a professor at the University of Southern California.

The work of Patrick Tighe Architecture has been published extensively, appearing in *Architectural Record, Architectural Digest, Global Architecture, The LA Times Magazine, Interior Design, LA Architect, Wallpaper, Metropolis, the New York Times*, and *Newsweek*. The firm's work has also been included in numerous architectural anthologies, including *The Power of Pro Bono* (Metropolis Books, 2010), and the latest edition of *The Phaidon Atlas of Contemporary World Architecture*.

Tighe lectures on his work and has been exhibited internationally. In 2013, the work of Patrick Tighe Architecture was included in the MOCA exhibit, *A New Sculpturalism, Contemporary Architecture* from Southern California. A monograph of the firm entitled Building Dichotomy will be released in 2015.

hitecture

Out Of Memory

Out of Memory, for the SCI-Arc Gallery in Los Angeles, was an experiential piece created through a convergence of sound, material, light, form, and technology. The installation was accompanied by a site-specific composition composed by world-renowned composer Ken Ueno. The soundscape was integral to the experience and used to explore the specialization of sound within the physical boundaries of the gallery. Memory was used as a vehicle to inform the work; the familiar was the impetus.

The installation proposed a new structural materiality through the use of renewable polyurethane foam. The foam was used as a complete building assembly: structure, envelope, and acoustic barrier. Layers of closed cell foam (used structurally) and open cell foam (used acoustically) were combined to create the wall assembly. The interior surface of the parabolic structure was a three-dimensional representation of the musical composition. Realized through onsite, six-axis robotic milling, the sonic contours were derived from the sound contained within. A spectrogram of the composition served as a source from which a mapping of frequency was translated into points and vectors. This provided a framework for the digitally modeled three-dimensional surface. The data was then used to robotically carve the interior surface of the volume.

Photography by Art Gray Photography

Twin Villa

The Twin villa is located 60 miles (97 kilometers) outside Beijing in the Pingu District of China in a new residential development surrounding a manmade lake. The Twin villa's dynamic form comprises a pair of monolithic volumes. The custom-faceted concrete shell structures are nestled into the sloping hillside, burrowing into the hill and cantilevering over the downslope reaching out toward the lake. By embedding the semi-subterranean building into the earth, the villa's impact on the landscape is reduced. The two wings function as the primary structural system supporting the villa's main circulation. A central circulation spine separates the pair of residential volumes. The interstitial space is a transparent threshold serving as a visual connector to the lake. Diffused light filters through the transparent void illuminating all levels of the home. As concrete is the predominant building material of the region and the preferred choice of the local craftsman, various applications of this versatile material are used throughout, exploiting its potential. Notions of fluidity, porosity, thickness, and smoothness are explored with regard to the relationship between the interior surfaces and the exterior building volume.

Photography by Patrick Tighe Architecture

Tigertail

The 3,200-square-foot (297-square-meter) residence is located in the Crestwood Hills neighborhood of Brentwood, California, a postwar development of modest mid-century homes. The design response is driven by pragmatic aspects of the topography and more complex three-dimensional manipulations of form. The architecture is a direct result of the various conditions inherent within the site. The peculiar geometry of the second-story volume is a result of the existing site conditions and the current building setback regulations. Openings to the views and the need for solid walls for shear and privacy were also factors that defined the building envelope. The building is low and unassuming at street level and in keeping with the scale of the neighboring homes. The residence opens to the courtyard with walls of glass. A series of bent, steel moment frames straddle the existing one-story structure and are expressed in the building. The folded planes of the walls and roof are an extension of the rolling topography of the hillside site. Views are framed as the building projects outward to the city, the ocean, and the neighboring Getty Center in the distance.

Photography by Art Gray Photography

La Brea Affordable Housing

The La Brea mixed-use affordable housing project for people living with disabilities is a 50,000-square-foot (4,645-square-meter) building for the city of West Hollywood. The building maximizes density while allowing for ample outdoor space. Thirty-two apartments are arranged around a shared exterior courtyard. Parking is provided at grade and commercial space fronts La Brea Avenue. The southwest corner is expressed as a beacon of activity and houses the circulation and other shared amenities for the residents. By nature, an affordable housing project is based on a simple, rational scheme. In this project, the design of the corner stands in stark contrast to other parts of the project. It is more expressive and more sophisticated architecturally, a counterpoint to the rational order of the affordable housing project, built with sophisticated robotic fabrication techniques and digital design. In reality, a fifth elevation, the corner is highly visible from Santa Monica Boulevard and La Brea Avenue. It serves as the entry, contains social spaces, gardens, and circulation. La Brea was built for the city of West Hollywood (through a nonprofit developer) to address an affordable housing shortage for tenants living with disabilities. The mixed-use program brings higher density to the urban core of the city.

Photography by Art Gray Photography

Todd Snapp
Principal

330 North Wabash
Suite 3600
Chicago
IL 60611
United States

www.perkinswill.com
todd.snapp@perkinswill.com
+1 (312) 755-4645

Every design problem is unique, and with my colleagues at Perkins+Will, the process of developing an idea into a unique response starts with no preconceptions. We begin with a blank slate, and develop an understanding of culture, the users, history, function, technical parameters, and many other sources of information that can influence and inspire an idea.

Throughout my career, not only as an architect, but also as a musician, carpenter, and contractor, I have always been infatuated with the art of assembling. The iterative process of assembling notes until they form a song, or the compounding process of assembling raw materials to form a building is fascinating. I tend to oscillate between two basic ways of approaching the assembly of a design idea, linear and progressive or casual yet relational. Maintaining a critical and iterative dialog between these two approaches, and employing a variety of tools to do so, is what I believe is the most important process for arriving at an innovative and meaningful solution wherever in the world a project is being designed and built.

The three projects submitted represent some important lessons I have learned during the course of my career at Perkins+Will. Each project is unique with respect to the overall context of the design problem, as well as the opportunities I have had to hone a process of both design and collaboration.

The Kennedy Expressway study explored sustainability and urban interventions in the form of an ideas exhibition at the Chicago Architecture Foundation. Rethinking proposed zoning and the possibilities of reconnecting two disparate sides of the city inspired a process that combined the practical problems of implementation and environmental sensibility with the fantastical ideas of stripping away the constraints of a specific site and letting the boundaries of the design evolve as the idea matured.

Perkins

The Tianjin History Museum submission synthesized a rich study of historical and cultural influences at a variety of scales throughout the design. The organizational framework and master plan recall significant aspects of the city's history, while the architectural development of the façade explored an algorithmic assembly of expressive components to evoke a character uniquely identifiable to the context of the museum.

The University of Angola, one of the first projects I worked on with Perkins+Will, is one that stands out for its innovative solutions, collaboration, and culture. The context of the problem demanded a culturally and sustainably responsive solution. The scale of the project was such that I had the opportunity to focus on key elements of the design and learn from extremely dedicated and experienced project leaders; and, the design solution was one of the most innovative that I have had the opportunity to collaborate on.

s+Will

Tianjin Museum

This museum was designed as part of a design competition for the master plan of a new museum campus for the city of Tianjin. The concept of the museum was based on an abstraction of a series of historical elements, which were key to the founding of Tianjin. The water and pedestrian walkways that surround the museum recall the saltern beds—part of the salt industry, which was, and still is, one of the major industries of Tianjin. The crystalline form of salt is also the inspiration for the façade system and atrium. The Crystal Atrium forms the grand preface hall and rises above the lower scale massing of the main building as a focal point. The crystalline salt is also the inspiration for the façade system, which consists of individually cast translucent glass cubes that will react with natural lighting conditions to enliven the presence of the museum within its context.

The museum's expression is also based on Tianjin's association with bridges, which provide adaptable methods of passage and are part of its urban structure. The uplifted form of the entry recalls the first modern operable bridge, which was located in Tianjin. The bridges in the atrium also recall the traditional pedestrian bridges in the historical structure of Tianjin. The galleries are organized around the central atrium, similar to the courtyard organization of traditional Chinese architecture, so that the visitor can interact with the exhibits thematically or chronologically. The window on the top floor is a focal point for the exhibition sequence, and also provides views of the old city along with the Tianjin City plaza below.

Photography by Todd Snapp/Crystal CG

A Green Corridor for the Kennedy Expressway

This project proposes an extended conceptual framework for the Kennedy Expressway corridor, west of downtown Chicago, based on the ideas put forth in the Central Area Plan (CAP) and Chicago Zoning Reform. The CAP proposes a park that bridges over the expressway in a one-block area between Madison and Monroe Streets, and the zoning reform suggests an increase in buildable area along this corridor, which would allow taller buildings to be developed.

This study expands the area to include the whole corridor between the tunnel known as Hubbard's Cave and the I-290 Interchange, visualizing a new green corridor for the West Loop. It also addresses the quality of development and degree of density along this corridor. The area consists of a series of park-covered bridges, which would link future developments on both sides of the expressway and stitch together the area of the city disrupted by the road. Verendeel trusses support green space above and museums, city offices, and retail within their structural depth.

The design of the parks and surrounding buildings explores issues of sustainability at a large scale. Extensively planted parks control and filter carbon dioxide emissions from the automobiles below, while the geometry of the bridges and buildings funnels clean air into pedestrian areas of the parks. The network of green continues vertically through a series of gardens incorporated in the high-rise developments.

Preserving existing and establishing new view corridors, from areas above as well as below the parks, is integral to the design. The intent is not to create yet another canyon, but to create a dynamic green link between downtown and the expanding neighborhood to the west.

Photography by Todd Snapp

Universidade Agostinho Neto

This project is the first phase of a university campus on the outskirts of Luanda, Angola, and illustrates how low-maintenance and low-energy sustainable urbanism can be applied to developing countries. The project consists of a central library with above and below-ground components surrounded by classroom buildings housing chemistry, mathematics, physics, and computer sciences. To reduce energy and maintenance costs, only the library is air conditioned while classrooms are naturally ventilated.

To maximize cross ventilation, bars are simple concrete structures with short east–west façades and long north–south façades facing the prevailing winds off the Atlantic Ocean. Over these bars is a louvered roof element whose shape and angle we calibrated through computer fluid dynamics modelling to create positive–negative pressure differentiation across the bars, maximizing airflow similar to an airplane wing. The roof also covers portions of the courtyards between the bars, creating a series of outdoor shaded gathering spaces. These bars surround a sunken library courtyard punctuated by a raised bar demarcating the central campus and housing offices. Once the central campus is finished in the second phase, the master plan calls for a series of specialized colleges organized along a series of pinwheel promenades.

Photography by James Steinkamp

Jennifer Marmon
Founding Partner

2404 Wilshire Boulevard, Suite 9E
Los Angeles
CA 90057
United States

www.p-ar.com
info@p-ar.com
+1 (323) 525-0990

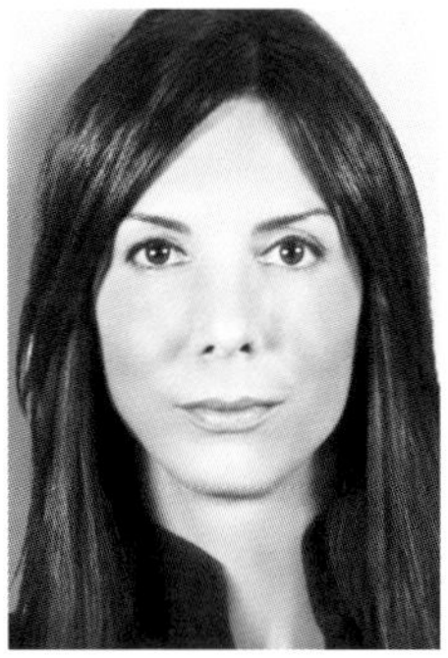

Platform for Architecture + Research (PAR) is an idea-driven office, committed to intellectual and artistic rigor and recognized for architectural innovation. Comprised of a collaborative team working on projects ranging in scale from housing and interiors to large, complex building schemes, PAR approaches each project with a fresh perspective, believing that each design arises from conditions particular to the site and program.

The work of PAR has received numerous awards and honors, including the American Institute of Architects Presidential Emerging Practice Award and an International Architecture Award in 2014 as well as an American Institute of Architects Honor Award and an American Architecture Award in 2013. Most recently, PAR has received international recognition in exhibitions at the Istanbul Architecture Biennale, the European Centre for Architecture, the US National Building Museum, the New York Center for Architecture and the Buenoa Aires Architecture Biennale. Our work has also received attention in publications such as DAMDI, Architecture, Detail, Future Arquitecturas and MARK.

Platform for Architectur

e + Research

Helsinki Central Library

In contrast to the other buildings in the Töölönlahti district of Helsinki, an essential component of our design involved creating a public space at the top of the library—visually connecting Töölönlahti to Senate Square and the city at large. The library is organized by six intersecting axes which afford spectacular vistas toward a celebrated landmark.

At the earliest stage of design, environmental analysis via simulation tools enabled design decisions that optimized the sustainable performance of the building form and envelope. The concave geometry optically magnifies light and heat, focusing it into the building and reducing energy demand.

A series of 38 foot (11.5 meter) deep floor plates are stacked to create a narrow building which optimizes natural light, thereby creating an ideal reading environment crucial for the library. The program of the library is subject to ever changing ways of accessing and sharing knowledge in a realm that is becoming more and more and immaterial.

The long span truss system provides a flexible, column free plan suitable to the demands of a 21st-century library and guarantees its viability as a building adaptable to future alternatives.

The landscape block is at the same time integrated and outstanding, traditional and contemporary. It creates a stage-like space for the annual gathering on the National Day of Finland, becoming a symbolic center for the city.

Photography by Labtop

Taichung Cultural Center

The new Taichung Cultural Center (TCC), located in the Taichung Gateway district, will house two regional agencies—the public library and fine arts museum—within one single institution. Our proposal arranges the new cultural center around a public space open to the city of Taichung and Gateway Park. The building is conceived as a single loop of public space and cultural institutions twisting into a continuous organization which combines the virtues of both institutions, maximizing interdisciplinary exchange while preserving autonomy.

TCC is conceived as a symbiosis of urbanity and nature. Like Taichung, which is located in the heart of the Taiwan mainland, it will be integrated into the heart of the park. The two institutions and their shared public facilities are gathered around an outdoor space framing a fourth program—an urban plaza which opens towards both Park Avenue 2 and Gateway Park. The multiple ramps and stairs of the building create an institution that is publicly accessible and welcoming on the inside as well as the outside. The urban plaza will attract the everyday life of Taichung flowing through its gateway while framing views of Taiwan Tower. Informal roof seating and stairs will make the TCC a lively place and a natural extension of life within the park. On special occasions it will turn into an outdoor gallery or urban stage to extend the art into the city, as well as the city into the institution.

The main entrance to the new Taichung Cultural Center lies to the north-west, adjacent to an auto drop-off along Park Avenue 2. Visitors enter through a shared lobby underneath the building's 20-foot (6-meter) lift, which leads to the public plaza. The south wing of the building tilts into the park landscape for continuous pedestrian access over the top of the building from Gateway Park and Taiwan Tower.

Photography by Luxigon

Lemesos Museum of Art

In 2012 the City of Lemesos held a competition for architects to propose designs for the Lemesos Museum of Art in the shoreline district. One of the competition's aims was to contribute to the consolidation of a cultural center composed of iconic architectural and programmatic attractants, projecting a singular and contemporary identity on behalf of the city.

Our proposal integrates and reveals the complex layers of history and urbanism that weave through this site, where the Lemesos Museum of Art will be situated for years to come.

Acknowledging these complexities, the strategy was to adopt the existing, diagonal circulation axis which generated a binary museum organization of permanent and temporary collections. Interstitial perimeter zones create a buffer between institution and city while affording impromptu events and large scale installations a visibility beyond the walls of the museum.

We were interested in offering an architectural vision where the paradox and conflicts between building and city, diversity and simplicity, and individual and populace erode. As Lemesos seeks to reinvent its cultural center, this building capitalizes on the seduction and impact of the singular image, while sustaining that effect through the attenuation of complex informational, structural, and atmospheric systems.

Photography by Platform for Architecture + Research

Hannes Koch
Founder

Florian Ortkrass
Founder

The Old Warehouse
2 Michael Road
London SW6 2AD
United Kingdom

www.random-international.com
studio@random-international.com
+44 (20) 7371-8064

Random International is a collaborative studio for experimental practice within contemporary art. Taking science as a means to develop a new material vocabulary, our work invites consideration of the man/machine relationship through explorations of behaviour and natural phenomena—with the viewer an active participant.

Random International

m
tional

Future Self

The installation Future Self presents viewers with their full-length body image, three-dimensionally distributed as tiny points of light. Quoting the minimalist tradition, we have reduced the image to the lowest resolution in which the reflected form can still be recognized.

Future Self studies human movement: what it reveals about our identity, and the relationship we have with our own image and with others. A slight delay in the depiction offers an encounter with the self-image that is direct yet ethereal.

This installation is extensively explored in a contemporary dance piece choreographed by Wayne McGregor and scored by Max Richter.

Photography by Random International

Rain Room

Rain Room is a 1075-square-foot (100-square-meter) field of falling water through which it is possible to walk, trusting that a path can be navigated, without being drenched in the process.

As you progress through the space, the sound of water and a suggestion of moisture fill the air, before you are confronted by a carefully choreographed downpour which responds to your movements and presence.

Photography by Random International

Todd Saunders
Owner and Principal

Vestre Torggate 22
5015 Bergen
Norway

www.saunders.no
post@saunders.no
+47 97 52 57 61

Our design philosophy is mainly about making exclamation marks in natural landscapes, while at the same time preserving them. I grew up in Canada and practice in Scandinavia. Therefore, materiality is mostly timber tradition and a Nordic material palette which we always use. But the forms are completely unconventional because we don't copy what's been done before. So we are experimenting with new forms all the time. And the work has more of a sculptural approach, as well as a functional approach. Those two things come together and result in our projects.

Saunde
Archite

ers

ecture

Fogo Island Inn

The Fogo Island Inn was conceived by the Shorefast Foundation, a Canadian charitable organization established by Zita, Anthony, and Alan Cobb. It is a building for learning that has emerged from four centuries of lived experience on the northeast coast of Newfoundland—to help carry the past into the future. It was created to be a cultural and economic engine for Fogo Island, one of Canada's oldest settlements; created in response to a pressing need to find new relevance for traditional knowledge and traditional ways. The goal was to "find new ways with old things." Fogo Islanders are a people who by virtue of their centuries of geographic isolation have become masters of making things by hand, recycling, and devising local solutions to all manner of challenges. Engaging this cultural and intellectual heritage was a key priority for us in the design of the building and was a key asset in its construction. The Inn is owned by a charitable foundation and is operated for the benefit of the communities of Fogo Island and Change Islands.

The main building is X-shaped in plan, with the two-story west-to-east volume containing public spaces, while the four-story southwest to northeast volume, parallel to the coast, contains 29 guest rooms. All guest rooms face the ocean and look onto the fishing grounds that attracted people to this island. The room sizes vary from 320 square feet (30 square meters) to 1075 square feet (100 square meters). The rooms on the third and fourth floors all having a wood-burning stove. The ceilings of the rooms on the fourth floor follow the slope of the roof and the three rooms on the east are double volume spaces with the sleeping area located on the mezzanine. Public areas include an art gallery curated by Fogo Island Arts; a dining room, bar, and lounge, which was recently rated as one of the top 10 new restaurants in Canada by *Enroute* magazine; and a heritage library which is home to the private collection of the late Dr Leslie Harris, former president of Memorial University of Newfoundland. The second floor includes a cinema that is a partnership with the National Film Board of Canada. The fourth floor roof deck has saunas and outdoor hot tubs with views of the sea. Traditional style "shore" legs are used to support the floors while minimizing the overall building footprint and the impact on the adjacent rocks, lichens, and berries.

We integrated ecological and self-sustaining systems subtly from the beginning of the project, incorporating technologies to reduce and conserve energy and water usage. The inn is a highly insulated steel frame building and the windows have the equivalent rating of triple pane glazing. Rainwater from the roof is collected into two cisterns in the basement, filtered, and used for the toilet and laundry water and also to be used as a heat sink for all of the kitchen appliances. Solar thermal panels supply hot water to the in-floor radiant heating as well as the laundry and kitchen equipment. The sound transmission classification of 69 between guest rooms ensures that guests hear only the sound of the nearby waves. The outbuilding houses wood fired boilers and solar thermal panels on the roof. The required number and orientation of the solar panels dictated the form of the outbuilding. The space between these two buildings creates an entry court and frames the main entrance. Vehicle parking is off site.

The knowledge and skill of local carpenters and craftspeople were essential for establishing the materials, details, furniture, and textiles used throughout the buildings. Their know-how was the starting point for what has become a long term and ongoing collaborative project between contemporary designers from North America and Europe and the men and women makers and builders of Fogo Island and Change Islands. The inn therefore serves as a means to reweave this remote island into the fabric of the larger world and its highly specific design has enabled it to contribute to the continuation of a traditional economy of care, craft, and culture.

Photography by Iwan Baan (left and far left); Alex Fradkin (above left)

Fogo Studios

Fogo Island, with a population 2,500, lies about 12.5 miles (20 kilometers) north-east of the island of Newfoundland, the vast and scarcely populated easternmost region of a vast and scarcely populated nation. Fogo Island is home to a gentle, independent people who have lived for centuries between wind and waves in pursuit of fish. The people are subtle and unpretentious yet have seen their traditional way of life in fishing threatened by forces largely beyond their control.

In 2001 Zita Cobb, a native Newfoundlander, began her own struggle against Fogo's creeping indigence and seemingly unarrestable depopulation. As successful entrepreneur, her idea was simple: to apply the same principles of investment-based development to Fogo and its communities, in an attempt to rebuild its economy by transforming it into a rural epicenter for artistic production of an international calibre.

The creation of a foundation and the architectural infrastructure for an artist residency program, according to Cobb's plan, would attract well-known artists; the international visibility derived from the program would bring high-end tourists; tourism would create jobs, and the island's economy would rebound. It is a process Cobb refers to as "social entrepreneurship."

Cobb opted for a masterplan—as opposed to a single design. This represents a key feature of the definition of the project's objectives: the default manifestation of the arts centre, i.e. a single, iconic, predictably monumental edifice, was ruled out from the start. In its place, six sites scattered across the island were identified; each would host a portion of the art centre's programs, allowing the artists and its resident guests to exist as a capillary network tightly interwoven with the daily lives of the local community.

It was clear to us from the outset that a particular architectural sensibility is required to engage a social and geographical ecology as fragile and severe as Fogo's. The initial assignment the architect received was to design six studios for artists and writers in residence, ranging in size from 215 square feet to 1300 square feet (20 square meters to 120 square meters), while traditional Newfoundland homes in various communities will be restored to become artists' residences.

The first of these studios, the Long Studio (opposite page), was completed in 2010. No road leads here—a 10-minute walk from the point where the nearest track ends ensures utter isolation, both physical and mental.

The proportions of the studio are linear, and project their occupant towards the Atlantic via an obliquely slanted, full height window, which frames the horizon or the shoreline, depending on one's position in the building. The intention was to sample or allude to local construction techniques: the pine-wood shell that cites the clapboard of the "outports," or local fishermen's houses; the pile construction of Newfoundland's waterfront huts; the proportions of the volumes.

There is a risk, in a context of such pronounced isolation, that angular geometries such as those that define the Long Studio might appear as gratuitous formal gestures; instead, the overall simplicity of the concept—a spare, open-space interior overlooking the shoreline, plus a semi-enclosed terrace area protected from the sea breeze, enclosed in the elementary form of a box—results instead in an architectural entity that is both sober and dynamic, bold yet not estranged from its setting.

All studios are entirely off the power grid with no connection to public services. In the Long Studio, all heat is produced from solar panels on the roof and a small wood stove, rainwater is collected from the roof and stored in tanks in the concealed storage rooms which ultimately supply water for the shower and small kitchen. In addition, a compost toilet is installed and all excessive gray water is treated inside the building.

All six studios, including Squish Studio (below), are autonomous—they do not rely on municipal water supply, sewer, natural gas, electrical power grid, or similar utility services.

Photography by Bent René Synnevaag

H. Koon Wee
Principal Architect

Eunice Seng, PhD
Principal Architect

Darren Zhou
Principal Architect

88 Luban Road,
Tower 4, Unit 301,
Huangpu District,
Shanghai,
China

158 Connaught Road West,
Unit 7A,
Hong Kong,
China

www.skewcollaborative.com
info@skewcollaborative.com
CN: +86 (21) 64319153
HK: +85 (25) 3960945

SKEW Collaborative is an architectural design and research practice founded by Singaporeans in the city of New York in 1999. The practice has been based in Shanghai and Hong Kong since 2007. Facing the need to reconcile these triangulated geographies, the cultural disparities between the cities become the raw material that feeds our studio's work. Through careful analyses of cultural phenomena and the natural and built environment, we seek to create architecture that is not only relevant, but also innovative and sustainable. More critically, the collaborative aspires to produce thought-provoking designs that can speak to the very condition of the discipline, within a larger theoretical discourse.

Acknowledging the contingencies in professional practice, our collaborative also actively participates in lectures, and art and architectural festivals. The collaborative has garnered numerous accolades for innovative work, including the LEAF Awards for Best Sustainable Development, Special Jury Prize for Best Residential at the Asia Pacific Interior Design Award, Best Mixed-Use and Best Sustainable Building at the Perspective Global Awards, Green Planet Top-50, Architectural Digest Top-100 Awards, and others. Projects were also previously shortlisted for the Best Sustainable Project and Best Small Project at the Blueprint Magazine Awards, World Interior News Awards, World Architecture News Awards, and LEAF ID Awards.

SKEW Co

In each project, we interrogate specific architectural artifacts in order to design architectural strategies that can bridge, critique, and translate between cultures and systems. To this end, our practice has amassed an archive of research pertaining to artifacts that include party walls, landscape patches, carcasses, industrial cores, curtain walls, and others. These artifacts are dissected, reconfigured, and recombined to produce opportunities for architectural frisson, for design innovation, or for rethinking the brief.

At the core of our practice and methodology is a feedback loop of professional, research, and academic platforms. As design principals we have taught at Columbia, Yale, and Princeton Universities, New York University, and Pratt Institute. We currently teach at the University of Hong Kong, leading core courses in history and theory, urbanism, and architectural design. Eunice is currently the Chair of the Ph.D. (DPRC) Program aimed at strengthening research methods and pedagogies in architecture. Koon and Darren direct the HKU Cities in Asia Summer Program designed to analyze and document complex processes of urbanization in Asia.

aborative

Jia Little Exhibition Center and Ateliers

This project integrates display and production spaces into a mixed-use complex, comprising an exhibition hall and three work/live atelier buildings. Through reinventing relationships between key constituencies, we seek to address both socio-economic and environmental sustainability.

The exhibition hall is a wrapped horizontal core that circulates through the buildings. The core is stretched, weaving visitors through different spatial experiences. We, therefore, re-thought the experience of the exhibition through this project, so that visitors are reconnected with the knowledge of back-of-house production.

The massing results in smaller zones for social interaction between the buildings, creating a micro-urbanism of the site. The courtyards become social condensers used by workers from the neighborhood, while the bridges allow for unexpected encounters through the spaces.

A key sustainable aspect of the project is the timber curtain wall of the exhibition hall, which responds to the building's requirements of sun-shading and views, while the directionality of the system promotes the legibility of the building as a horizontal core. The atelier buildings are contrasted with the highly expressive hall, with vertical window systems designed to provide sufficient daylighting while keeping a tight thermal envelope.

Photography by Cai Feng

Wulumuqi Road Roof House

The project starts with a beautification policy in Shanghai where pitched roofs are added to buildings. The original flat roof was demolished to give way to an attic and references the triangular dormer windows found in the neighborhood.

The triangulation is extrapolated to give shape to the ceilings, walls, and staircase. Lighting is concealed along the edges of overlapping triangles, while the staircase takes on a similar form, where each tread and baluster marks the culmination of these surfaces. The new continuous form gives an organizational logic to other programs, such as a small outdoor roof patio, and a passively cooled interior, where hot air rises and escapes.

Many of the original elements, such as the wood floor and steel windows, are retained with a maximized use of recycled building materials. Clay tiles, bricks and timber beams came from demolished neighboring sites. Supplementary heating systems are placed along the perimeters to minimize heat loss and gain.

Local builders using traditional techniques were hired during construction. This project therefore connects to the socio-economic roots of its surroundings by searching out the old generation of craftsmen who rebuild with recycled materials from a city undergoing rapid demolition.

Photography by Shen Qiang

Xishuidong Industrial Heritage Retail District

The project comprises five new buildings in a cruciform retail street configuration, framed alongside two clusters of historic industrial buildings. The design scope includes a series of low-density retail architecture linking several landscaped public plazas, and a bridge.

The site sits on a canal intersection and is surrounded by heritage industrial buildings and signifies the future, which anticipates changes from an industrial to a livable city.

In preserving the original fabric, new street-level architecture was stitched between the historic buildings, so that one can fully experience both new and old architecture, while avoiding the typical podium-and-tower commercial typology. This reduces the dependency on air-conditioned spaces and allows for different ways of relating to the gated residential community and newly created piazzas and bridges.

The new architecture redeploys the north light roof forms found onsite, to bring about a more humanized scale and rhythm in the experience of the street while maintaining the availability of daylight in the new retail spaces.

The traditional industrial materials found in existing buildings are transformed into new applications such as sun louvers, screens, terraces, lighting surfaces, commercial displays, and even urban furniture.

Photography by Alvaro Quintanilla / Iuliana Chiras

Florian Idenburg
Principal

Jing Liu
Principal

Ilias Papageorgiou
Principal

68 Jay Street, Suite 501
Brooklyn
New York 11201
United States

www.so-il.org
office@so-il.org
+1 (718) 624-6666

SO–IL is an idea-driven design office that brings together extensive experience from the fields of architecture, academia, and the arts. A creative catalyst involved in all scales and stages of the architectural process, SO–IL approaches projects with an intellectual and artistic rigor fueled by a strong commitment to realizing ideas in the world. SO–IL is led by partners Florian Idenburg, Jing Liu, and Ilias Papageorgiou.

We have worked on an array of projects ranging in scale from the master plan of a cultural campus in Shanghai, China to a series of prints for the Guggenheim Museum. Projects include the flagship store for Benetton in NYC, student housing in Athens, Greece, the Shrem Museum of Art at the University of California in Davis, the Frieze Art Fair in NYC, and Kukje Gallery in Seoul.

Our firm's work has received numerous awards, including the MoMAPS1 Young Architects Program, the AIA Young Practices Award and the Emerging Voices award by the Architectural League. SO–IL has been widely featured in international publications such as *The New York Times*, *The Wall Street Journal*, *Fast Company* and *Domus*, and in exhibitions at the Guggenheim Museum and the Museum of Modern Art in New York, the Los Angeles Forum for Architecture and Urbanism, Studio-X in Beijing, Kunsthal KAdE in The Netherlands and the Benaki Museum in Greece. The work of SO–IL is part of the collections of the MoMA, the Art Institute Chicago, and the Guggenheim Museum.

SO–IL

K3 Gallery

For Kukje Gallery in Seoul, SO–IL has developed a master plan of the gallery's "art campus" in the historic urban fabric of Sogyeok-dong, a low-rise area in the northern part of Seoul. Small alleyways and courtyard houses characterize this neighborhood, which is currently being infiltrated by newly constructed galleries, boutiques, and coffee shops. One of the newly planned buildings within the master plan is SO–IL's 11,000-square-foot (1,022-square-meter) gallery. Considering the clear diagrammatic geometry of the white cube too rigid within the historic fabric, the building is enveloped in a permanent "nebula"—a pliable chainmail veil. The stainless-steel mesh produces a layer of diffusion in front of the actual building mass, through a combination of multidirectional reflection, openness, and the *moiré* pattern generated through interplay of its shadows. It is strong yet pliable, and can easily wrap around crude geometries. The structure is a single-story, clear-span art space. A perimeter skylight admits and diffuses natural light. The ground floor is used for large installations, performances and other functions, while the two sub-level floors house a 60-seat dark stained wood-clad auditorium, administrative areas, catering spaces, restrooms, and mechanical spaces.

Photography by Iwan Baan

Frieze Art Fair

Frieze Art Fair is an international contemporary art fair that takes place every October in London's Regent's Park and since 2012 in New York every May. The fair is staged by Amanda Sharp and Matthew Slotover, the publishers of *frieze* magazine.

Frieze Art Fair features more than 170 contemporary art galleries, and the fair also includes specially commissioned artists' projects, a talks program and an artist-led education schedule. Working with a prefabricated rental structure to house the inaugural New York presence, challenged SO–IL to be inventive with a limited vocabulary. Randall's Island, situated in between Manhattan, Bronx, and Queens, is one of the few pieces of open space in New York City large enough to accommodate the 225,000-square-foot (20,900-square-meter), 1,500-foot-long (457-meter-long) structure.

Pie-shaped wedges are inserted between six tent sections to open up the standardized system, and offer amenities at each section as a moment of relief. The wedges divide the relentless length of the fair and bend the otherwise straight tent into a meandering, supple shape. The winding form animates the visitor experience both inside and out. It also establishes the temporary structure as an icon along the water.

Photography by Iwan Baan

Logan HQ

Logan, a bi-coastal video production company, commissioned SO–IL to design a new office space for their New York HQ. We reconceived how office space could work given Logan's organization, creating layers of transparency and completely shared workspaces.

The space is divided into two identical, symmetrical rectilinear spaces. Each long room has a 65-foot (20-meter) continuous custom worktable. Accommodating working groups of any size, the shared desks consolidate almost every operation–design, production, and meetings–of the company in one place. End sections of each table are divided by glass walls, allowing for acoustically private offices and meeting rooms while sharing the same work surface. Seamless floor-to-ceiling translucent fabric walls separate the central work areas, visually compressing the scale of the space, while maintaining a shared environment, and allowing natural light to penetrate.

A stretched PVC luminous ceiling provides even, shadowless lighting.

Looking through layers of fabric, people and objects perpetually appear out of focus. Fabric walls catch natural light and change colors as light changes throughout the day. This abstraction, combined with the symmetry of spaces, the spatial ambiguity of the fabric, creates the illusion of reflections when looking into adjacent spaces, creating a dream-like, surreal experience.

Photography by Iwan Baan

Alvin Huang
Principal

724 S. Spring Street #1101
Los Angeles
CA 90014
United States

www.synthesis-dna.com
alvin@synthesis-dna.com
+1 (213) 430-9967

It can be argued that the catalyst for every major advancement within the discipline of architecture has been a parallel advancement of technology. These technological advancements vary from technique (the invention of perspective by Brunelleschi), to material process (the mass production of steel utilized by Paxton in the Crystal Palace, London), to building technology (the invention of the elevator by Otis).

More recently, the introduction of computation within the field of architecture has led to several advancements within our discipline. These advancements encompass all three previously described categories, as computation has enabled new generative design techniques, new digital fabrication processes, and new performative building technologies.

Our work explores the creative application of advances in computational design research through the lens of a forward-thinking and multi-award-winning contemporary architectural practice in search of new architectural paradigms. Our body of work spans all scales, including buildings, façades, interiors, pavilions, furniture, and objects.

It is through the overlay of design research and design practice that each of our projects is explored to test ways in which computational processes can inform and augment the design, development, and making of progressive, socially engaged, and sustainable architecture.

Synthesis Design + A

rchitecture

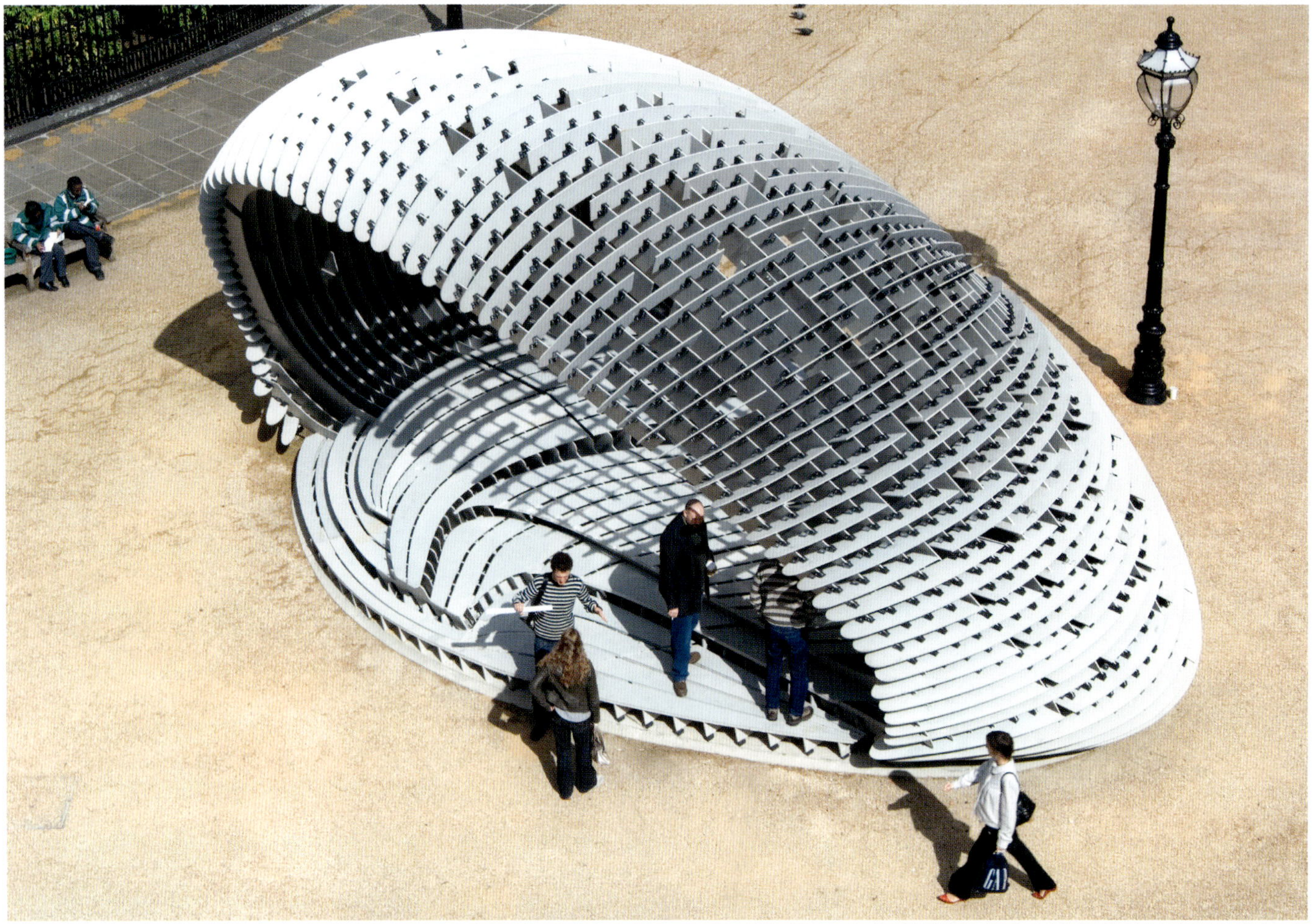

Digital Craft Experiments: [C]Space Pavilion & Chelsea Workspace

Our [C]Space project was the winning entry in the AADRL10 Pavilion Competition. It is an advanced technology concrete structure that was erected in Bedford Square, London in March 2008 to commemorate the 10-year anniversary of the Design Research Laboratory Graduate Design Program at the Architectural Association.

The pavilion is formed by a discontinuous shell structure spanning over 33 feet (10 meters) made of thin fiber reinforced concrete elements which perform as structure and skin, floor walls, and furniture. The concrete is manufactured by the Austrian company Rieder, and the design of the pavilion takes the material to new technical limits, which required extensive prototyping and material testing during the design development process. The jointing of discrete concrete profiles exploits the tensile strength of Fibre-C and a simple intersecting notch joint which is locked together using a bespoke rubber gasket assembly. The angle of intersection at each joint continuously varies across the structure. We conducted the entire design process using 3D digital and physical modeling, and we completed the design development using rigorous constraint modeling and scripting to control over 850 individually different profiles and 2,000 joints. Finally, the elements were manufactured directly from digital models on CNC cutting equipment using standard sized 0.5-inch (13-millimeter) thick flat sheets of Fibre-C concrete and 0.6-inch (15-millimeter) thick mild steel plate.

Chelsea Workspace is a home office which responds to the client's brief of a discrete yet sculptural design solution for a home office by conceptually draping a dynamic surface over an arrangement of the required home office elements, including a work desk, storage units, printer, and paper shredder.

Set within a tiny 75-square-foot (7-square-meter) room in the client's flat, our design intention was to minimize the imposing presence of the boxy storage requirements by disguising them within a sinuous form that emerges from the walls of the room. As the owner desperately wanted to reduce the visual clutter of "stuff" in his work space, we concealed all storage for books, filing cabinets, printer, shredder, electrical cabling, phone, internet, and lighting controls within the volume.

The space is organized around the lone window in the room, in order to provide the client with a pleasant view and ample natural daylighting. The glow of natural light from the window further serves to articulate the geometry of the installation through the play of light and shadow.

The fluid geometric form is articulated as a series of alternating CNC milled birch plywood ribs, which are prefabricated into modular units and spray finished with a clear lacquer. These units are hung on a series of Z-clips bolted to the interior wall, and held in place by under their own weight.

We arranged horizontal spacers, which stabilize the open ends of the plywood ribs, in a pixelized graphic pattern representing a world map, where the owner (a seasoned world traveler) can map out his travels.

Our project is the result of an intense collaboration between designer and fabricator. This dialog was made possible through the exchange of 3D files not only as design deliverables, but as production and fabrication documents.

The result merges the digital with the handcrafted to produce a stunning result.

Photography by Valerie Bennett (top); Phil Rogers (bottom left); Peter Guenzel (bottom right)

Form Finding Experiments: Pure Tension Pavilion and Minimal Relaxation Installation

Our two projects represent a series of applied design research projects exploring the use of parametric physics solvers to simulate the physical properties of tensioned membrane structures through a computational process known as Dynamic Mesh Relaxation.

The Pure Tension Pavilion is a portable, solar-powered tensile membrane structure commissioned by Volvo Car Italia for the new Volvo V60 plug-in hybrid electric car, which not only charges the car but also flat-packs to fit in the trunk of the car and assembles in less than one hour. It is an experimental structure which, similar to a concept car, is a working prototype which speculates on the future of personal mobility and alternative energy sources, while also exploring digital design methodologies and innovative structural solutions.

The pavilion is an expression of the tensioned equilibrium between its elastic membrane skin and rigid perimeter frame. The perimeter frame of the structure is defined by 24 CNC bent aluminum pipes with swaged slip fit connections, while the skin is materialized as a pair of vinyl encapsulated polyester mesh membranes with a zippered seam and spandex sleeves which wrap the frame. The entire pavilion collapses neatly into two 65 x 15 x 15-inch (165 x 38 x 38-centimeter) "B-cases" and weighs 154 pounds (70 kilograms) in total and can be assembled and taken down in just under an hour by a team of two to three people.

Some 252 lightweight, flexible photovoltaic (PV) panels are embedded in an applied graphic pattern of vinyl tiles along the mesh. The PV pattern is the result of intensive solar incidence analysis on the structure, which found the average annual solar incidence of the skin for 360 degrees of orientation and mapped PV locations to the areas of greatest average annual solar incidence. Additionally, we utilized a Maximum Power Point Tracking controller to sample the output of the cells, and selectively disable those that are not collecting enough energy, thereby ensuring that the pavilion is receiving as much charge as possible. The pavilion can recharge a fully depleted car in about 12 hours in optimum sun conditions.

Minimal Relaxation is a temporary roof canopy and landscape installed on the roof terrace of the MOCA Shanghai. The installation reinterprets the Chinese garden to activate the MOCA roof terrace as an undulating and responsive landscape that simultaneously produces gradient spatial conditions and framed viewing portals which curate views of the surrounding high-rise towers. This canopy examines the potential of gravity forces to derive geometry by relaxing a net, in which every web member reveals its load path, expressing its catenary form in multiple dimensions. The final installation was realized through a series of tensioned net structures and shrink-wrapped skins. We embedded the undulating landscape with color-changing LEDs connected to motion sensors, with color-changing responses to express intensities of occupation.

The highly successful installation activated the previously under-utilized roof terrace into a popular summer evening destination. It also became a civic landmark viewable from surrounding high-rise towers.

Photography by Mosca Partners (top and bottom left); Wandile Kraai (bottom right)

centralworld
KRUNGSRI

The Groove @ Central World Expansion

The Groove @ CentralWorld is a 130,000-square-foot (12,000-square-meter) extension to a 6.5 million-square-foot (600,000-square-meter) shopping center in Bangkok, Thailand. Situated in the center of retail artery of Ploenchit Road, the building sits proudly at the intersection of CPN Tower (the Thai developers' corporate headquarters), and CentralWorld (the world's third largest mall). As such, our design of the relatively small development serves a much larger urban (and brand) purpose as both icon and entry point into CPN's corporate and retail empire. However, as a spatial experience, the project attempts to provide an intimate oasis for dining, entertainment, and boutique shopping as a departure point from the three conditions it mediates: the intense traffic of Ploenchit Road, the corporate empire of CPN, and the huge crowds of Central World.

Our design does this by using the massing of the building to integrate three ubiquitous urban elements of Bangkok's urban condition into the project: the courtyard house, the covered terrace, and the Soi (pedestrian alley). These three conditions are manifested to create a sequence of open air, naturally ventilated, and shaded spaces for walking, gathering, and dining with integrated landscaping and water features.

Capturing the vitality and intensity of Ploenchit Road's pulsing traffic, our design of the anodized aluminum composite façade is characterized as a series of curving and angled flows articulated with varying intensities of backlit perforations. As the façade peels away to expose recessed storefronts and covered terraces, solid teak trim, soffits, and suspended ceilings are revealed at all horizontal surfaces to express the inner flesh of the warm and intimate interior of the project.

Designed as a lightweight steel structure to sit atop the existing below grade reinforced concrete parking structure, we designed the entire fast-track project in six months and it was constructed in just six months.

Photography by Polar Factory Photography

Melkan Gürsel Tabanlioğlu
Partner

Mesrutiyet Caddesi
67 Tepebasi
Beyoglu
34430 Istanbul
Turkey

www.tabanlioglu.com
info@tabanlioglu.com
+90 (212) 2512111

As relationships are central to our existence, yet one of the most difficult areas of human life, starting a project I always focus on the question "who would live here and what would be the lifestyle?" The most important thing in life and in practice is to make people happy, and for that people should enjoy their surroundings; the places they live, work, and entertain. Searching for new efficiencies in terms of global and environmental needs and developments, my works comprise a wide range of building types from housing, offices, mix-use, and industrial projects, to high-rises and public facilities.

I believe in research and a team approach. The team includes engineers and consultants, and the client of course. Respecting relations with the environment in all senses, open to probabilities and possibilities including technological innovations and especially smart local solutions, we search for the unique and the right answers and bring solutions while manifesting our architectural language. With reference to vernacular wisdom and state-of-the-art technologies at the same time, we develop projects for today and the future, referring to existing values and capacities

TABANL
Architec

Architecture is a system and you have to collaborate and adapt to the situation for the best and most feasible result. If a building works well and the people who use it directly and even who only pass by are happy, it is a good building. For making better cities today, we need to overlay different activities in the same space; we need to repair the collectivity of space and re-build the ceremonial and informal public spaces of our age. To motivate an interactive, lively social mood, we need to offer multifunctional, flexible architectural approaches, which will define the new social routines towards a more extrovert urban lifestyle. The transitivity and synergy born out of architecture will encourage such a transformation. The new community centers and extension of the public zones will be acquired though transparency and openness.

Every person has the right to achieve contemporary living standards and have a joyous life. The construction sector depends greatly on politics and economics, and in every case I believe we have to use our abilities to demand and bring about a more equal world. Hence, our architectural contribution is important to motivate the vision of the governors, and make people aware that a better quality of life is possible.

Our buildings provide a medium and signify the will to create a more transparent and interactive society.

Tabanlioğlu Architects was established by Murat Tabanlioğlu (RIBA Chartered, AIA Int.) in collaboration with Dr. Hayati Tabanlioğlu in 1990. Melkan Gürsel Tabanlioğlu (AIA Int.) joined the group in 1995. Based in Istanbul with branches in Ankara, Dubai, Doha and London, the office has more than 160 employees, operating mainly in Turkey, MENA and CIS Countries. Tabanlioğlu has more than 60 years' experience in architecture, searching for new efficiencies in terms of global and environmental needs and developments. We have been awarded several international and national prizes, including 2013 Mipim AR Future Project Awards, Big Urban Project for Yenikapi Transfer Point and Archaeo-Park Area project, 2015 MİPİM Award for Selçuk Ecza Headquarters, RIBA International Award (Royal Institute of British Architects) for the Loft Gardens, for the Bodrum International Airport and many others. The company ranked 66th "BD World Architecture Top 100 list" in 2014.

IOGLU

ts

SIPOPO Congress Center

The Congress Center is situated by the ocean and is surrounded by the woods. The lacy texture of the mesh behaves like a sun-break and grants a shady interior without hindering the accent of transparency. Protecting the inner glass walls reduces this way improves the building's energy efficiency. A double-space atrium lobby grants a feeling of spaciousness, and creates a wide circulation zone. The main conference hall is a rectangular space at the first floor, with a circular seating design, situated amid the structure and accessed through the connecting lobbies at three sides. The restaurant overlooking the bay is at the first floor flanked by the foyer.

Photography by Emre Dörter

Tripoli Congress Center

Open to the world cultures and where diverse languages meet up, Tripoli Congress Center is a strong element of high representation and welcomes presidents of the world in the texture of the natural environment. Surrounded by the woods, the rectangular two-story "block" is nested in a semi-transparent metal envelop as a "shield" that opens up to the external landscape with a wide portico that defines the main entrance. Between the mesh and the inner glass walls, a semi-open shady circulation area is created as an extension of the landscape. This provides natural cooling. Incise patterns inspired by the trees around the site permit controlled daylight to diffuse into the central space and at night the oozing beams through the slits integrate with the wood at the background. It is simple but functional and finds its expression in clarity and sincerity.

Photography by Cemal Emden

Loft Gardens

Following the success of first Levent Loft, a transformation project on Maslak-Levent axis, Loft Gardens is designed in the concept of "soft loft." The 21-story, mid-rise, transparent residence block comprised of loft apartments of various size and forms; nature friendly with its vertical gardens designed as private walkout terraces, balconies, and garden patios. The loft apartments have floor-to-ceiling, one-piece, energy efficient windows opening to the city panorama. A wealth of natural light creates more commodious and ambient spaces. Exposed ductwork, concrete flooring, masonry and corrugated steel elements are kept in the spirit of openness and creativity that are integral to loft design. Yet, the apartments have defined spaces, especially bedrooms, fine flooring and they feature high end kitchen and bathroom fixtures and finishes whilst the key elements of openness and versatility in loft design are respected.

Photography by Helene Binet

Tatiana Bilbao
Principal

Paseo de la Reforma 382-301
06600 Mexico City DF
Mexico

www.tatianabilbao.com
g.alvarez@tatianabilbao.com
+52 (55) 85898822

We practice architecture to improve people's quality of life. We work with our environment and the materials that surround us, using manual labor and other techniques, opening channels of communication between different social groups, and carry out productive activities that enable different aesthetic experiences with strong ideas and direct definitions and intentions.

We try to understand our world, through our multicultural and multi-disciplinary office, in order to translate its rigid codes into architecture. Through these strands, the office regenerates "humanized" spaces which respond to global capitalism, opening up niches for cultural and economic development.

As architects we create a climate of collaboration where there are various disciplinary resonances in technical areas, and theoretical and artistic works which we believe can in one way or another affect the patterns and structures of society.

The office associates work to the theme of resonance, which matches the frequency of a given system with the frequency of an external drive, with certain information generated by another system.

As with the ethics of otherness according to Levinas, the office incorporates the other, which has not been recognized or accepted by the intellectual, political, and business élite, but which is on the lookout for a qualitative change and structural life. At the end we are building with the responsibility of understanding all that we do and we mean to do: we learn though it and we work with it.

Tatiana

n Bilbao

Casa Ajijic

The client requested an adaptable, integrated, vernacular, and low maintenance summerhouse. To achieve this we combined a simple housing program with the geometry that responds to the way it is constructed.

The cubes of the house represent the single parts of the family, which exist as individuals but also as part of the whole. Constructing with the materials that surround the site helped us to rediscover an ancient system which we often forget, preoccupied as we are with the latest technology. Colors and textures in the walls and the landscape merge to bring house and environment together.

Photography by Iwan Baan

Botanical Garden Culiacán

The Botanical Garden is a place where nature, art, and architecture coexist with the intention of offering a new place to learn and improve the quality of life within a contemporary cultural environment. We developed a masterplan to integrate landscape, botanics, and art, including services that allow this new program to merge and give a complete reference to all visitors.

The educational compound in the Botanical Garden in Culiacán is formed of three separate buildings; an educational room for kids and teachers that can host workshops; a 100-seat auditorium for different kinds of screening and lectures; and a service building with toilet facilities.

We conceived the small open air structure in the Botanical Garden to be a place for up to 70 people where a seven-minute video tour of the botanical garden will be screened as an introduction to the visit. This space with walls on three sides is covered only by the shadow of the surrounding trees and offers fresh air to those who enter it.

Photography by Iwan Baan

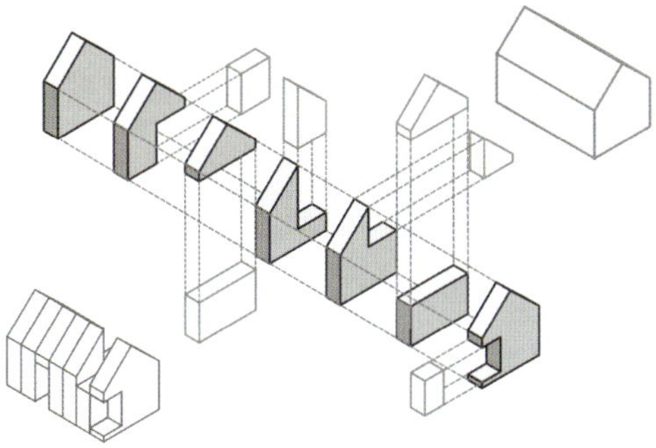

Ordos 100 / House A

The concept for this project emerges from the study of seven different sections of the archetypical house. Each section undergoes a process of addition and subtraction to generate the final solid volume.

The core concept of these seven sections creates a continuity of exterior and interior spaces which correspond to each other. Our design creates spaces that transform the inside-outside relationship and develop a form that questions the extroverted and introverted nature of a house.

Bolle Tham
Co-founder and Chief Architect

Martin Videgård
Co-founder and Chief Architect

Blekingegatan 46
11662 Stockholm
Sweden

www.tvark.se
info@tvark.se
+46 8 702 00 46

Architecture is about the future. A building stands for hundreds of years, so inevitably it is not only contemporary but also a form of continued history projected into the future. Seen from this perspective, all the ordinary circumstances affecting the conception of a new building or a new urban environment are secondary to the lasting effect and quality of the construction: its long term environmental effect, its architectural integrity and function, and its capacity to convey the ideas and ideals of society. The end result is the only thing that matters.

One factor that makes architectural work constantly interesting and inspiring is context. If one looks closely enough at a project's context—its physical, cultural, and ideological environment—there will be an inexhaustible source of inspiration for new solutions to any architectural program. In fact, one can argue that it is primarily the approach to context that determines whether an architectural project works—both at its location and within its own time. It follows that architecture should be closely allied to its contexts and be firmly anchored to earn its place in reality. Originating from our interest in multiple contexts, this is an investigation we pursue in our practice and in every project we undertake. It has resulted in an approach where the analysis of a site and the interpretation of a functional program can produce an architecture that is diverse, integrated with and informed by its specific place in society, whether in wild nature, or in a city.

When looking at architectural work we often return to a discussion on what we call "the operative elements of architecture." All great works of architecture—be it the Mayan pyramids, La Rotonda, the Lingotto, the Villa Savoye, or the Kunsthaal—have strong cultural presence and offer distinct spatial experiences. These constructions are all consequences of their specific cultural and physical context but also results of a precise and exact handling of space and how it is

Tham & Videgård

organized. Spatial relationships, light, movement, proportion, scale, mass/void, rhythm, repetition, and structure are the cornerstones of the experience of architecture. These concepts—the operative elements—are the architect's primary tools. They are the building blocks that shape the actual experience, and the effect is consistent, regardless of epoch. We live in a time when, through real and virtual travel, the world is getting smaller.

Globalization provides people around the world with common references, presenting architects with new opportunities—a more open attitude toward different ideas—but also challenges—the risk of neglecting context and of reproducing meaningless uniformity. However, a combination of analysis of the physical and cultural context and precision in the use of the operative elements of architecture enables one to generate site specific and distinct buildings, an architecture that is relevant and meaningful. Basically, it is about separating architecture from style and instead beginning with perception, physical impact, and lived experience. The best architecture is clear enough to offer multiple readings depending on whom, when, and where you are.

Tham & Videgård Arkitekter are based in Stockholm, Sweden, and directed by co-founders and chief architects Bolle Tham (b.1970) and Martin Videgård (b.1968). Since the start in 1999 the practice has been awarded prizes and nominations in several international competitions and has also attracted attention for its experimental and innovative projects. In 2005, Tham & Videgård won, from among 294 proposals, the open international competition for the new Kalmar Museum of Art. It was awarded the Kasper Salin Prize 2008 for best new architecture in Sweden. It was also shortlisted for the Mies van der Rohe Award 2009; the European Union Prize for Contemporary Architecture. In 2007, after prequalification of four international teams, including Swedish Wingårdhs and Japanese SANAA, Tham & Videgård won the competition for the new School of Architecture and the new Campus Entrance at the Royal Institute of Technology (KTH) in Stockholm, opening in 2015.

Major works include: the new School of Architecture, the Kalmar Museum of Art, the Moderna Museet Malmö (for the Swedish Museum of Modern Art), the Tellus Nursery School, the Uppsala Departmentstore, the Archipelago house, the House Lagnö, the Creek House, the Apartment Humlegården and the Tree Hotel in Harads.

The office has also gained opportunities to work abroad and is regularly invited to participate in international exhibitions. Most recently at a solo exhibition at La Galerie d´Architecture in Paris, the Lisbon Triennale, the 12th and 13th International Venice Architecture Biennales, the Victoria & Albert Museum in London UK and the Louisiana Museum of Modern Art in Denmark.

This year construction starts of a new school building at the Krabbesholm School of Art, Architecture and Design in Skive, Denmark, the result of a competition 2014.

As an integrated part of the architectural practice, Bolle Tham and Martin Videgård also teach and lecture at schools of architecture in Sweden and abroad. Recent lectures by T&V were held at the Netherlands Architecture Institute, Rotterdam (NL), at ETH Zürich (CH), École Spéciale d'Architecture and the Pavillion de l'Arsenal, Paris (FR), at the Architectural Association (AA) School of Architecture, London (UK), at the Royal Academy of Arts, London (UK), at the Harvard Graduate School of Design (GSD) in Cambridge and at the IIT, the Illinois Institute of Technology, Chicago (USA).

Kalmar Museum of Art

The new Kalmar Museum of Art is set amongst the high trees in the main park of the renaissance town of Kalmar, an unusually dense cultural and historical context. It includes part of the remains of the medieval city wall and the 1930`s restaurant pavilion by Swedish modernist Sven Ivar Lind.

Kalmar is a flat town and the museum creates a new vertical profile that complements the horizontal condition of its urban landscape. Four floors are stacked on top of one another creating a vertical ascent up into the greenery of the trees, giving a series of different spatial experiences while offering views of the surroundings; Kalmar castle, the lake and the city centre. A new place in the midst of a very discernible historical context that presents new perspectives on contemporary society.

The competition motto was "Platform", also the conceptual idea of this museum: a series of open platforms for art related activities. The construction utilised large spans for maximum flexibility on each level, so that not only light but also space can be transformed and adjusted to meet the specific needs of each exhibition. Almost domestic in scale this museum still provides a variety of exhibition conditions.

Kalmar Museum of Art won first prize in the 2004-2008 Open international competition.

Photography by Åke E´son Lindman

Tree Hotel

In the little village of Harads, close to the Arctic Circle, the brief was to design a small cabin - a place to spend the night up among the treetops. Reflecting on the remote and beautiful site chosen for the project and the way we generally encounter the elements and explore the wilderness, there arises an intriguing paradox. The eagerness to experience the most extreme conditions, such as the challenges of a harsh climate, wildlife and difficult terrain; co-existing with the desire for total protection and control, which means being outfitted with highly engineered materials and high-tech equipment (Gore-Tex, Kevlar, carbon fiber, super-light tents, etc.)

Inspired by these observations, the Mirrorcube blends in completely with its surroundings but also stands out as an abstract piece. It is invisible and a landmark at the same time. It is built with a lightweight aluminium structure, anchored to the trunk of a tree and sheathed in a skin of mirrored glass. The glass reflects the forest all around, minimising the structure's visual impact in nature, creating a camouflaged refuge among the trees. From inside there are wide-open views of the surrounding environment. A hidden roof terrace is accessed from inside.

Photography by Åke E:son Lindman

Creek House

On the south western coast of Sweden, in an old agricultural area, the site is located on a gentle slope leading down to the sea. An exceptional plot in the midst of the cultivated, open landscape as it is planted with many different trees and flowering bushes, endowing it with the character of a small botanical garden. Large trees stand close together and the sun streams through the leafy canopy with a small creek meandering through the garden further enhancing the idyllic setting.

The home is spread out over a single story and each function is contained in a separate spatial unit making the program conspicuous with boxes of varying scales and heights, constructed from handmade bricks of a deep red colour. The complex of spaces cannot all be viewed from a single point and they thus appear larger and quite enigmatic and secretive.

Building and its site each make visible qualities in the other, like the small creek that passes through the house, bridged over by the entrance hall. Inside, a few steps separate common and private areas into different levels and each room is characterised by a vaulted ceiling giving spatial direction in relation to the landscape outside.

Photography by Åke E:son Lindman

Tim Durfee
Principal

639 South Spring Street, 4A
Los Angeles
CA 90014
United States

www.timdurfee.com
tim.durfee@gmail.com
+1 (213) 840-1820

Everything is connected. At least that is what the telecom ad says. This extreme linkage of all things is profoundly transformative to how our lives are lived: altering not only how we experience the world, but even how we "know." My project is to operate in these connections, the new inbetween. I grew up in Pittsburgh (the Brussels of the USA), and Brussels (the Pittsburgh of Europe). I come from a family of seven: an artist, a banker, a physicist, an architect, a journalist, an athlete, and an engineer. I have always been drawn to subjects about everything: literature, architecture, and media.

I studied English in college. Two things I learned from literature:1. The best work has clear form and the form itself is the idea; 2. The best writing is—in the words of Karl Ove Knausgaard—"not what happens there, not what actions are played out there, but the there itself." Both of these things are also true of architecture.

Architecture school at Yale exposed me to an exhilarating cacophony of design cultures. It taught me to bristle at the question, "What kind of architecture do you like?" "Good architecture," is the only imaginable answer. Siloed groups or styles only motivate me more to trangress them.

During my years teaching and directing the Visual Studies program at the Southern California Institute of Architecture (SCI-Arc), I came to understand that technology's most lasting effect on design will not be solely the expansion of formal possibilities, but in the expansion of design itself.

Tim Durf

Before forming Tim Durfee Studio, I created exhibitions for major museums. Exhibitions are information spaces: architecture whose form emerges from the mandate of communicating ideas and information itself. As we face a future where information is only more and more pervasive, I believe that the best exhibition designs provide clues to the architecture of that future.

Several years ago I joined the Media Design Practices MFA program at Art Center College of Design as the "architect" in the small faculty, which is curated to represent different perspectives on design (along with a technologist, a filmmaker, a graphic designer, an anthropologist, and a social researcher). We re-conceived, re-structured, and re-named the program to create a new type of advanced design degree to address new contexts, mediums, and methods in design.

At Art Center College of Design, I founded amp: a studio for multi-modal projects exploring the changing relevancy of architecture in the age of media. amp might be an acronym for "Architecture and Media Programs," or "Amazing Mediated Peripheries," or "Architecture, Maybe? Precisely!"

With amp and Tim Durfee Studio (and with frequent collaborations with Iris Anna Regn), I am pursuing the ambition of creating a new role for the architect and designer. One who views architecture as a continuum through all scales and mediums, including—but not limited to—buildings themselves. We find ourselves in a world saturated with stories and images and information. It may be invisible, but it is transforming our very existence. It demands the reimagining of design itself.

I seek to make work that answers questions one never thought to ask. Work that addresses overlooked forms of experience, context and material. Work that responds to how people live now, not 20 years ago. Work that feels discovered, not invented. Work that has a hard-earned clarity (I believe Occam can sometimes use a chainsaw, sometimes an X-Acto). Work that is sincere, never earnest. Funny, never frivolous. In an era of likes and pokes and tweets, I believe striving for timelessness is the new avant garde.

ee Studio

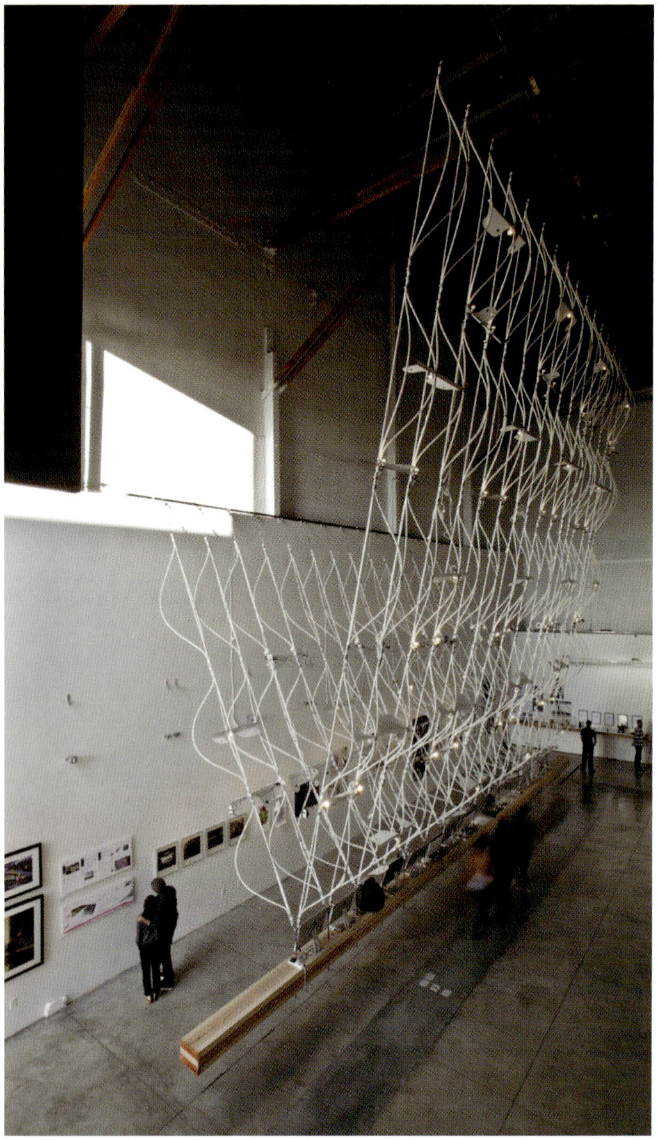

The Rather Large Array

This 49 x 49-foot (15 x 15-meter) suspended building-sized scanner was fabricated to both present and record the exhibition *MADE UP: Design's Fictions* (curated by Tim Durfee with Haelim Paek) in the Wind Tunnel, a gallery hosted by the graduate Media Design Practices (MDP) program at Art Center College of Design. The structure explores a possible future for architecture where sensors, cameras, and other digital technologies are so central to the functioning of our cities that they become one and the same with the physical architecture.

As visitors view the work in the gallery, the structure periodically scans the space with its 24 cameras. The images are then compiled and plotted onto the exhibition guide, producing a "live" catalog of the exhibition (one in which visitors often find themselves among the work.) The installation is comprised of 256 thermally formed PVC pods of three strands each: one for data, one for power, and one for structure. The designers worked with Buro Happold to engineer a one metric tonne suspended wooden beam, which itself supports 14 video displays.

While the purpose and technology of the Rather Large Array are utterly different from its inspiration—the famous Very Large Array, constructed in the 1970s in New Mexico—the underlying principles are similar. Unlike the Very Large Array and other landmarks of Big Science, however, the Rather Large Array uses only modest hardware store components: inexpensive microcontrollers, cameras, and lights. The construction approach allowed for nearly all of the material to be reused: few of the fourteen 21.5-foot (6.5-meter) Douglas Fir beams were cut and could be reused; the steel rigging materials were rented for the installation period; and since no holes compromised the integrity of the 768 PVC pipes, they could be donated to a non-profit housing organization and urban gardening project.

Photography by Joshua White

Growth Table

Few activities are as universal as drawing. For nearly every young child, drawing is a joyful compulsion. For the rest of us—even those un-inclined artistically—we share memories of the youthful pleasures of just sitting down and making something.

In a public space, the Growth Table creates an instant intergenerational community by welcoming people of all sizes (including those still crawling, at the very bottom) and providing easy access to tools and surfaces for drawing.

Tim Durfee & Iris Anna Regn
Photography by Jeremy Eichenbaum

L.A-Frame House

The L.A-Frame House employs an often maligned typology to respond to the conditions of its site—and proposes an alternative to the prevailing strategies of residential development in Los Angeles.

Mount Washington, like many residential areas of Los Angeles, is suburban in character but urban in context.

Increasingly, however, the impulse to build available lots to the maximum allowable volume is altering the nature of this once unique form of LA neighborhood. Specifically, the new development tends to eliminate the visibility of private backyards from the streets which, in a city impoverished of public green space, is often the only form of visual and spatial relief.

The L.A.-Frame House instead stacks its program vertically into a pavilion, and provides a deck to augment the living and playing area of the backyard. Occupants thus travel "under the yard"—past a workspace and library—to get from one structure to the other. This unusual path of interior circulation also enables clear visibility of the mountain view from the entire site.

While the origins of the form were non-referential, once the A-frame profile emerged in the development of the design it was embraced as an appropriate foil; the existing 1954 cabin was a product of the same *piney* post-war culture of leisure which built the original California A-frame houses, and the juxtaposition of the old and new structures can serve to represent, in their modest way, a debate between the expansive ethic of LA's horizontal past, and the possible benefits of a more vertical future.

Photography by Tim Durfee and Iris Anna Regn

Tetsuya Matsui
Architect

1-3-28-102 Midorigaoka
Ikeda
Osaka 563-0026
Japan

www.uzu-a.com
uzu@uzu-a.com
+81 (72) 747 6780

Architecture originated to protect people from the rigor of nature, such as rain, wind, and snow. It has grown remarkably as a shelter with the passage of time, but at the same time the relationship with the natural environment has become poor. I endeavor to keep architecture in contact with people's natural surroundings.

To think about architecture is to think about the relationship between people and the relationship between people and nature, beyond time and space. And it is to create the circumstances which are based on these relations. An architect needs to listen to many different voices and accommodate them carefully. It is necessary to respond properly to everything.

UZU Arc

hitects

Arima Guest House

The Arima Guest House is located at the corner of the street, where a prostitute is said to have lived many years ago, in an old spring hot resort called "Arima". Forty-five years have passed since this renovated common house was built after the Second World War. It features tiled roofs, mortared outside walls, a bases of clay walls, plaster walls with cotton-wool fiber and glass sliding screens with a wooden frame. Tall, old maple trees in the neighboring property are frame the approach. At night stars shine above the trees.

Our client offers the guest house as a personal place to stay for his many friends in other countries. We would like the guests to relax in the atmosphere with history and nature in this place to keep a structure of the former Japanese residence. We left wooden poles and removed ceiling boards to show the existing structure such as wall bases in the ceiling without drastic change of shape and painted dirty and cracked walls as if they put on light make-up, leaving the feeling of original materials. Finally we installed nets made of linen with light and shadow generating there on the wooden poles and at the back of paper sliding screens.

We adjust the old style to the new style, and appreciate the harmony. We wanted to express something like the trace of a prostitute who turned into the breeze like rustling maple leaves many years later.

Photography by Akiyoshi Fukuzawa

Cross House

There are two mountains in front of the Cross House. The two mountains can be seen from each facing wing of the house, which is designed with Japanese-style timberwork and colored walls in French style. It is located near a village at the foot of Mt Hiei, from which water flows down.

Our client is a French chef who cooks creatively by using local ingredients, while his Japanese wife loves making things.

The house is a combination of the chef's sensitivity and enthusiasm and a mixture of France and Japan brought to life in natural materials.

We painted parts of the timberwork, installed lighting equipment, laid decking boards, made a tank for holding rainwater, and assembled a kitchen.

The kitchen is in an arched shape to allow views of the two mountains from both inside and outside the house, and designed over split levels to accommodate a slope on the plot.

Photography by Akiyoshi Fukuzawa

Otonoha Preschool

Otonoha Preschool is a nursery school that provides children with natural surroundings and music. The school's symbol is a tree with a musical note. Rice paddies are to be found all around the city. In the surroundings, we designed the floors so they connect to the rice paddies outside, and the hills where children can climb and look at the scenery.

The children become familiar with how rice, the Japanese staple food, is grown. The children one year old and younger see the scenery of farming outside the window from the room on the first floor everyday. The children from two years old to five years old experience farming with the farmers in the neighborhood. They plant rice, watch living creatures, harvest rice, make rice balls with their parents, and have lunch parties.

Children feel the rhythm when they play eurythmics and they also feel the sound when they run together with it. The room on the second floor can be divided from one room to twelve spaces by wooden partitions and lace curtains with various sizes and atmosphere. The deck in the south connects to the big stairs and the green slope where a school teacher planted clovers and the field. This is a space that has an image for the practice of eurythmics. The inside and the outside are connected by a mixture of various sounds such as the sound of instruments, the voice of children, the sound of wind in the fields, and insects.

We designed the school to add the softness of nature to the interior. The light and curtains harmonized with the motion of wind and children. Everything around the environment becomes as one.

Photography by Akiyoshi Fukuzawa

Pankaj Vir Gupta
Director

Christine Mueller Gupta
Director

C7/125
Safdarjung Development Area
110016 New Delhi
India

www.virmueller.com
christine@virmueller.com
+91 (11) 2656 5633

While we continue to speculate on the opportunities offered by architectural design, we are convinced that acts of spatial imagination must be tempered with equally seasoned aspirations for construction. Every component of architecture we imagine must then be drawn over several iterations in order to evolve an articulate template for construction. By creating projects with locally available materials and crafted by mainstream construction workers, vir.mueller architects assert the primacy of construction means and methods as the basis of a contemporary practice. Following stringent standards for drawing and detailing, we are eager to learn from our collaborators onsite—the brick mason, the carpenter, and the general contractor.

We believe in working closely with our consultants and in embracing sustainability as an engineering mandate in a manner that is inherently elemental. We have learnt that a well-considered and simple structural diagram may create provocative architecture in itself.

We strive to create timeless aesthetic relationships, evoking the spirit of contemporary culture, ecology, and technology.

According primacy to the process of design, our work evolves as a response to the particular character of each program, site, and budget.

vir.mue
archite

As humanists, the social and contextual background of our material choices is as important as their aesthetic value.

Designing and building thoughtfully are vital to the sustenance of our practice, if only as an antidote to the indifference with which the average Indian regards architecture. We are fearless about adopting anachronistic tendencies whose long-term benefits are considerable; hence our preference to build with materials expressed in their natural state. We take great pride when our designs empower daily wage construction workers to consider themselves artisans. Rather than bemoaning rapidly vanishing craft traditions, we are energized by the prospect of elevating standards of construction across the building industry.

The ethos of our studio is to perpetuate to as wide an audience as possible, an alternative means of experiencing architecture. We'd rather design for the many than for a select discerning few. For vir.mueller architects, the act of architecture is also an act of resistance, a refusal to cater to the accepted praxis of design and construction as perceived in India today. The idea of citizen architects, visible to the public as an informed voice of design-related issues, defines our identity as an architectural studio. This philosophy has guided our endeavors in our years of teaching and consulting for public and civic agencies.

We have been fortunate that our work has spanned a range of typologies, scale, and context. This has allowed us to redefine for ourselves how the verb to live—encompassing to work, to think, to eat, to play—is actually enacted in space.

Every project is an opportunity to explore the relationship between the culture of a place—both physical and material—and its craft traditions. The geographic context of our projects—Boston, Colorado, Udaipur, Delhi, Ahmedabad, Kolkata, Chandigarh—have offered us significant variations in the quality of light, the color of foliage, and the topography of landscape, making us aware of a more pliable sense of the elemental. By discovering the roots of a place through such cues, we participate in galvanizing this context.

eller

cts

Defence Colony Residence

This project resurrects and perpetuates an ancient architectural craft of load-bearing, brick masonry construction. Like all craft traditions, this practice represents a confluence of cultural, material, social, and economic practices. Brick has always been a fundamental building module in Indian architecture. Moreover, it links modern architecture with an ancient craft tradition, refined over centuries of use. In reviving this material, and especially using it so extensively—as structure, surface and detail—this project makes a strong and coherent argument for linking sustainability with this most basic and local of all building materials.

The unbroken lineage of brick masons, trained by their ancestors, risks being marginalized as construction practices in India are becoming increasingly mechanized. In a frenzy of appearing "contemporary," buildings are assembled from catalogs. The art of making architecture with a singular, basic material—local earthen brick—is considered anachronistic! As the new Indian architectural culture seeks international expression, this project expresses the vitality and beauty of building with a local, sustainable, and basic material. It also sustains the livelihood of a guild of brick masons struggling to retain their place in society.

Photography by André J. Fanthome

Delwara Community Toilets

The village of Delwara is a small, peri-urban settlement located 17.5 miles (28 kilometers) from Udaipur, Rajasthan. The villagers work primarily on small agricultural plots adjacent to their homes.

Several commute daily to Udaipur for work. Few of the villagers have access to toilets within their residence. The development of basic sanitation infrastructure such as plumbing and sewage disposal is absolutely critical for Delwara. This project sought to create the fundamental infrastructure required for a community toilet—the first of its kind in the town.

Given the scarcity of water in the desert and the high levels of ground water contamination from septic tanks, a dry composting prototype was selected as the optimal model for this site. Incorporating rubble masonry for walls, and a bamboo framing system for the roof, the design emphasizes an aesthetic of non-industrial materials crafted by local artisans. The scale of this project was determined by the limitation of semi-skilled and skilled craftsmen working entirely by hand—the villagers are skilled in stone masonry and carving and proposed to participate in the construction.

Our design sought to embed this project within the available technology and traditions of Delwara's citizens.

Photography by vir.mueller architects

Wolkem Office Building

Our design solution for the new office building for Wolkem India Ltd, in Udaipur, Rajasthan, emerged from simple principles—all employees must have direct access to natural light and ventilation, the materials used should require minimal maintenance, and the energy efficiency of the building should be maximized.

Our aesthetic vocabulary amplifies these principles. The east and west façades are designed as insulated brick masonry cavity walls, while the north and south elevations incorporate insulated glazing systems and deep balconies which are screened by sandstone louvers. The façades are clad in local sandstone, contrasting with the exposed concrete beams and slabs. The executive wing, sited in the north-east corner of the site, has a view of the desert hills.

Sited in an arid climate, the building remains grounded in a courtyard oasis—with trees casting dappled shadows on stone walls, and the sound of water echoing from a reflecting pool.

Photography by André J. Fanthome

Jury

In the 21st century, according to economics specialist Gerald S. Hobbs, "what worked yesterday, may not work today and will never work tomorrow." Flexibility is the key to achieving results, and this was certainly the case in the jury process for this volume of *Fifty Under Fifty: Innovators of the 21st Century*.

The invited participants, Stanley Tigerman, chair, partner Tigerman McCurry Architects, Ralph Johnson, Global Design Director and Principal, Perkins+Will, Jeanne Gang, Founder Gang Studio, Marion Weiss, Co-founder WEISS/MANFREDI and Qingyun Ma, Founder MADA s.p.a.m. are frequently in flight, bound for global clients and activities. Despite having three of the jurors based in Chicago, it still proved impossible to capture all five of them in one room on the same day .

When Tigerman and Johnson convened in the Perkins+Will Chicago office, Gang was in Australia and Ma in China. Weiss was in New York, although linked through voice conferencing.

Authors Russell and Ameen caught up with Dean Ma in his office at the University of Southern California, Los Angeles, to solicit his careful selections. Ms. Gang received the work of the 50 firms in a vast three-ring binder and gave her preferred choices over email. It was very apparent that despite the lack of verbal interaction they all concurred on the finalists.

Such a contradiction of traditional jury procedure can only be described as a typical example of 21st-century innovation, in synchronicity with this book.

Jeanne Gang

Studio Gang Architects

Founded by Jeanne Gang in 1997, Studio Gang Architects is an award-winning architecture, interiors, and urbanism practice based in Chicago and New York. Studio Gang helps organizations, communities, and cities achieve their mission and goals through solution-focused problem-solving and transformative architecture. A collective of architects, designers, and thinkers, the studio operates as a laboratory—identifying patterns; testing ideas, materials, and physical properties; and developing new models at various scales.

They begin each project with a research-based discovery process, accruing resources including material and historical precedents, ethnographic studies, and real-time data on environmental and urban conditions. This shared body of knowledge informs prospective concepts and is also used to test and refine the design, ensuring the final project resonates with its specific site and culture, while addressing urgent global themes, such as urbanization and climate. Iterative experimentation using both digital and analog technologies and frequent collaboration with renowned experts in a wide range of disciplines are likewise integral to the studio's process. Embedded in this way of working is the conscious, ongoing effort to reduce a project's energy and material footprint, simultaneously seeking out the exciting synergies made possible by sustainable design.

From cultural and educational institutions to communities and cities, clients play an integral role throughout the life of a project. Together with the client and stakeholders, Studio Gang achieves a deep understanding of the project's mission and goals, identifying and implementing innovative yet pragmatic strategies to achieve them.

Guided by this approach, Studio Gang has produced some of today's most compelling works, including the Arcus Center for Social Justice Leadership, Kalamazoo, Chicago's WMS Boathouse at Clark Park, the Aqua Tower, and the Nature Boardwalk at Lincoln Park Zoo. Recently, the studio was selected to design the expansion of the American Museum of Natural History in New York, and is currently engaged in major projects throughout the United States and Europe. These include the Fire Rescue 2 training facility in Brooklyn, New York, the Campus North Residence Hall at the University of Chicago, and several towers and large-scale buildings in San Francisco, Miami, and New York.

As well as creating transformative architecture, they are dedicated to employing the tools, methods, and visionary capability of design to catalyze positive change through public engagement and advocacy.

Jeanne Gang, FAIA, LEED AP

Architect and MacArthur Fellow Jeanne Gang, FAIA, is Founder and Principal of Studio Gang Architects. Internationally recognized for her innovative use of materials and environmentally sensitive approach, Jeanne explores the role of design in revitalizing cities. Through projects ranging in scale from community anchors and cultural institutions to tall mixed-use buildings and urban design, she engages pressing contemporary issues and their impact on human experience.

Her work has been honored and exhibited widely, including at the International Venice Biennale, the Museum of Modern Art, the National Building Museum, and the Art Institute of Chicago. A distinguished alumna of the Harvard University Graduate School of Design, she has taught at Harvard, Yale, Princeton, Rice, and IIT, where her studios have focused on cities, ecologies, materials, and technologies. Her published work includes *Reveal* (Princeton Architectural Press, 2011), her first volume on Studio Gang's work and process, followed by *Reverse Effect: Renewing Chicago's Waterways* (2011), which envisions a radically greener future for the Chicago River and Great Lakes.

Solar Carve Tower, New York, The form and mass of the tower benefit the public by allowing more light, air, and views to the shared amenity of the High Line Park.

Rendering courtesy of Studio Gang Architects

Arcus Center for Social Justice Leadership, Kalamazoo, Michigan, The Arcus Center offers a highly visible, inclusive space for social justice activism. The center embraces its three varied contexts—the Kalamazoo College campus, a residential neighborhood, and an old-growth grove—with transparent façades, encouraging a greater sense of community.

Photography by Steve Hall © Hedrich Blessing

Aqua Tower, Chicago, Aqua's terraces provide access to fresh air and views, and enhance social connectivity among neighbors by offering the potential to interact on the building's façade.

Photography by Steve Hall © Hedrich Blessing

Writers Theatre, Glencoe, Illinois, The theater's lobby and gallery provide a warm welcome for visitors to this important cultural anchor and can be transformed for various purposes, including informal performances, educational outreach, audience enrichment, and celebration events.

Rendering courtesy of Studio Gang Architects

Nature Boardwalk at Lincoln Park Zoo, Chicago, Conditions of the nature boardwalk vary with different seasons, local cycles of drought, and more global changes in climate. This variation challenges the aesthetic of manicured landscapes and offers educational and scientific opportunities to study these changes.

Photography by Steve Hall © Hedrich Blessing

Marion Weiss

WEISS/MANFREDI

Marion Weiss is cofounder of New York–based WEISS/MANFREDI Architecture/Landscape/Urbanism, and the Graham Chair Professor of Architecture at the University of Pennsylvania. A multidisciplinary firm, WEISS/MANFREDI operates at the nexus of architecture, art, landscape, and urban design. The firm's Olympic Sculpture Park, Seattle, exemplifies this cross-disciplinary design approach and the project has been recognized internationally through museum exhibitions and design awards. *Time* magazine identified the park as one of the top 10 projects in the world, Barcelona's World Architecture Festival selected the project as winner in the Nature Category, *i-D* magazine awarded it the highest Environment Design Award, and it was the first project in North America to win Harvard University's Veronica Rudge Green Prize in Urban Design.

The firm received the American Academy of Arts and Letters Award, which acknowledged the distinct vision of the firm and was named one of North America's "emerging voices" by the Architectural League of New York. The firm was also honored with the New York AIA Gold Medal and the Tau Sigma Delta Gold Medal, awarded to one individual or firm in the world each year. In 2014, Weiss was inducted into the National Academy of Design. The firm's projects have been featured in exhibitions at The Museum of Modern Art, the Venice Biennale, the São Paulo Biennale of Architecture and Design, the Cooper-Hewitt Museum, the National Building Museum, Harvard University, the International Landscape Architecture Biennale in Barcelona, and the Guggenheim Museum.

WEISS/MANFREDI recently opened the Krishna P. Singh Center for Nanotechnology, which was cited by the *Wall Street Journal* as one of the best buildings in the world, "going way beyond the basics of a research lab ... enough to persuade one that architecture really can prompt the kind of nonlinear thinking that leads to discovery." The projects, Taekwondo Park in South Korea, and Hunters Point South Waterfront Park and Brooklyn Botanic Garden Visitor Center, New York City, have all been recognized for the inventive weaving of architecture and landscape to create socially vibrant and ecologically resilient public destinations.

Recently, WEISS/MANFREDI also won the national competition for the National Sylvan Theater at the Washington Monument Grounds in Washington, D.C., which is conceived as a seamless extension of the National Mall's path system, offering a new multi-level sequence of routes into and through these historic landscapes. Other noted built works include the Barnard College Diana Center, the Museum of the Earth, the Women's Memorial and Education Center at Arlington National Cemetery, and the pro-bono Robin Hood Foundation in New York.

The firm's first monograph, *Site Specific: The Work of WEISS/MANFREDI Architects* (Princeton Architectural Press, 2000) precedes *Public Natures: Evolutionary Infrastructures* (Princeton Architectural Press, 2015), which envisions new reciprocities between infrastructural systems to create ecologically resilient territories for public life and features 10 recent projects by the firm.

Marion Weiss teaches design studios focused on the synthetic potential of infrastructural, social, and ecological challenges at the University of Pennsylvania, where she is currently the Graham Chair Professor of Architecture. She has also taught at Harvard University, Cornell University, and Yale University, where she earned her Master of Architecture degree.

Qingyun Ma

MADA s.p.a.m.

Founded by Qingyun Ma in 1996, in New York, and launched in Shanghai in 1999, MADA s.p.a.m. has evolved into the most energetic and creative design firm in China. Based on the firm's transcultural mission and cross-Pacific practice, it has become one of the most visible firms both in China and around the world, through its critical thinking and participative practice. MADA's work has been widely published and exhibited, including the first cultural facility exhibited in the Venice Biennale, the Centre Pompidou in Paris, and the first solo show from China in Aedes Gallery in Berlin. Currently, it has more than 50 designers and architects based in the head office in Shanghai.

It is MADA's belief that architecture is a collective endeavor and building execution is a team effort. The firm's strong team spirit is apparent in all of their projects, and quickly becomes a fundamental for leadership among their collaborators, subsequently, successes in many of projects. The quality of MADA's architecture reflects the quality of their teams—proactive, responsive and decisive.

Lead by Qingyun Ma, MADA's approach to architecture has been cross-territory and inter-disciplinary. Among which, creative participation in cultural infrastructure and commercial confluence has been the most noticeable. Ma has been curator in significant exhibitions involving governments, private sectors, and art circles. An affiliated company, SPAMAPS has invested and facilitated the first contemporary art museum and artist residence program in Xi'an, China. This approach to urbanism and architecture manifests itself in part of the firm's name "s.p.a.m.," which stands for strategy, planning, architecture, and media.

They approach a problem through broad-based investigation of causes and consequences, in order to arrive at the most effective and original way to reach a goal. Strategy is a merger of imagination and rationalization. It precedes all organizational and design efforts.

Masterplanning for them is an indiscriminative selection and orchestration between nature and the manmade, existing and emergent, big and small. It is a structure that stipulates parts without stiffening them. Through these juxtapositions, it simultaneously projects life onto a physical setting from outside, as well as invigorating life from within.

They see architecture as a process of selecting and configuring materials and technology in relation to finance, technology, and the idea. It is a construct of ideas as well as a product. They also see architecture as a form of knowledge gained through new and creative exploration and constant refinement.

Architecture is a form of representation. To extend and continue the attitude and proposition of architecture, media plays an important role. As our living environment becomes increasingly non-physical, architecture must re-form, or even de-form in order to accept media, and to define our environment in a way previously unseen.

Qingyun Ma

Named one of the world's most influential designers by *Businessweek*, Qingyun Ma's experience and involvement on an international level makes him a leader in the design community with a great understanding of contemporary issues in global urbanization.

Ma earned his bachelor's degree in Civil Engineering in Architecture from Tsinghua University in Beijing in 1988, and in 1991, he received his Master of Architecture from the University of Pennsylvania. Through his academic and professional experience, he developed a keen business sense centered on design intelligence, which culminated in his establishment of MADA s.p.a.m. in 1996. MADA s.p.a.m. has designed and built modern buildings and urban communities, garnering worldwide recognition. He also founded spamALL as a design incubator in Beijing's world-renowned 751 and co-founded GalleryALL for contemporary design in Los Angeles and Beijing.

Through his unique position within the discourse of urbanism and architecture, Ma has become a frequent participant in many exhibitions and publications. In 1996, he coordinated Rem Koolhaas' first "Harvard Project on Cities," which resulted in the book *Great Leap Forward/Harvard Design School Project on the City* (Taschen, 2001). He wrote the main curatorial text for a 2003 exhibition on Chinese contemporary art and architecture at the Centre Pompidou in Paris. He has curated and participated in numerous exhibitions, such as the 2004 *MADA on Site* traveling European exhibit, the 2007 Shenzhen and Hong Kong Bi-City Biennale on Urbanism/Architecture. He was a conceiver and curator of the exhibition *C.A.N. Project* (Culture, Agriculture and Nature) at the 2014 Venice Biennale. He also served as a planning expert to the International Olympic Committee for the 2008 Olympic Games in Beijing, and served as a juror on the prestigious Rome Prize and the 2010 Shanghai World Expo.

In 2007, he became Dean of the School of Architecture at the University of Southern California. As Dean, he set an aggressive agenda for expanding the scope and skillset of the students and faculty. His main focus is integrating technology and sustainability using the digital process. Also, by organizing workshops around the world, Dean Ma strives to establish USC as a world-class school of architecture.

Dean Ma has focused the school on his message of "global culture, creative future" by launching the USC American Academy in China (AAC). AAC serves as a base for visiting scholars to facilitate their research and foster academic exchange and is the only academic and cultural bridge between the United States and China encompassing architecture, arts and the humanities. It is an enduring program for American students and scholars and welcomes international student participation. Since its inception, AAC programs have fostered the exchange in the disciplines of arts and architecture in China through academic tours, discussion forums, and intense workshops surrounding urban design and planning, art, and architecture.

Qingyun Ma also established the first privately owned winery in China, Jade Valley Wine & Resort, which is now one of the most respectable Chinese wine brands.

Ralph Johnson

Perkins+Will

Since 1935, Perkins+Will has created innovative and award-winning designs for the world's most forward-thinking clients. They are architects, interior designers, urban designers, landscape architects, consultants, and branded environment experts who approach design from all scales and perspectives. Engaged, accessible, and collaborative, their staff of 1,500 professionals brings together high design, functional performance, and social responsibility to advance project goals. Inspired by the programs within, they design from the inside out. They combine a deeply humanistic approach with results-driven pragmatism to create dynamic spaces for people. Research-focused and inventive, the firm reimagines how space can be used to foster stronger ties between communities, the built environment, and nature. With nearly 1,000 LEED Accredited Professionals, sustainable design and the use of healthy building materials are fundamental to its process. Their transformative designs help students learn better, patients heal faster, business teams perform stronger, and city dwellers have more meaningful daily experiences.

Ralph Johnson, FAIA, LEED AP

Ralph E. Johnson received his Bachelor of Architecture from the University of Illinois, and his Master of Architecture from Harvard University. He began his career at Stanley Tigerman's office and then joined Perkins+Will in 1976, where he currently serves as its National Design Director and is a member of its Board of Directors. His projects have been honored with more than 70 design awards, including nine national Honor Awards and more than 50 regional Honor Awards from the American Institute of Architects, as well as a Progressive Architecture Design Award.

He has lectured at numerous universities and has been a Visiting Critic at the Illinois Institute of Technology, the University of Wisconsin at Milwaukee and the University of Illinois. Johnson is a member of the Board of Overseers, Illinois Institute of Technology, School of Architecture and the Committee on Design Architecture of the Art Institute of Chicago. He has been elected as a Fellow of the American Institute of Architects.

Universidade Agostinho Neto

Photography by Steinkamp Photography

Contemporaine

Photography by Steinkamp Photography

Johnson's designs have been exhibited at the Milan Triennale, the Art Institute of Chicago, the Paris Biennale, and the São Paulo Biennale. Monographs on his works were published by Rizzoli in 1995, L'Arca in 1998, and Oscar Riera Ojeda Publishers in 2013.

His recent work includes the Kenya Women and Children's Wellness Centre in Nairobi, Kenya, the Shanghai Nature Museum in Shanghai, China, a university center for Case Western Reserve University in Cleveland, Ohio, a laboratory for Massachusetts Institute of Technology, and the Athletics Center and Visitor Center for Northwestern University in Evanston, Illinois.

William Jones College Preparatory High School

Photography by Steinkamp Photography

Case Western Reserve University Tinkham Veale University Center

Photography by Steinkamp Photography

Stanley Tigerman

Tigerman McCurry Architects

Tigerman McCurry Architects has its roots in the Chicago tradition of innovative design. The firm, as it is presently constituted, is the result of the melding of two individual design practices that combine nearly eighty years of experience. Tigerman's practice was initially founded in 1962. McCurry's architectural and interiors practice was founded in 1977 following her 11-year tenure at Skidmore Owings and Merrill. In 1982 the two firms merged. Tigerman McCurry Architects' diverse portfolio includes institutional, educational and religious facilities, museum installations, mixed-use and affordable housing projects, retail showrooms, restaurants, recreational facilities, and private residences. Their geographical scope of work encompasses 17 states and 10 foreign countries, always operating as a small, personalized, professional office. Today it consists of its two Principal Architects, two Senior Associates, three intern architects, and administrative support.

Tigerman McCurry is the recipient of seven national American Institute of Architects Honor Awards and more than 150 national and local awards for architecture and design excellence. Our work has been exhibited worldwide over 300 times at museums and galleries, such as The Museum of Modern Art in New York, La Jolla, and Teheran, the National Museum of Modern Art in Kyoto, the Deutsches Architecktur Museum in Frankfurt, The Metropolitan Museum, Guggenheim Museum and The Art Institute of Chicago. The firm and its principals have been published nationally and internationally more than 4,200 times, including in a special edition of *Architecture + Urbanism* magazine depicting the complete works of the firm.

The two principals of Tigerman McCurry Architects, Stanley Tigerman and Margaret McCurry, both Fellows of the American Institute of Architects, are committed to the same goal: the creation of a contemporary and authentic American architecture that is characteristic of its own time and place. This philosophy leads them to assemble the standard ingredients essential to architectural production—to interact, to question, to challenge, to collaborate—in order to reach a consensus and to create a building that represents the collective energies of client and society. They embrace this hybridization as intrinsic to an American culture that is the inevitable result of the melting pot. The work of the two principals defines an architecture that sometimes reflects an obligation to contextual responsibilities, and sometimes strikes out individually to show a better way. Their ethical concerns are rooted in general practice, resisting the urge to become expert in one thing at the sacrifice of many things.

Tigerman McCurry Architects asserts a conscious effort to address issues surrounding social cause, environmental, and ecological concerns in their quest for design excellence. They inherently believe in designing environments that enhance the psychological and physiological wellbeing. Toward this end, the firm's principals were early advocates of incorporating sustainable design approaches and technologies into their work. Recent projects such as the International Masonry Institute National Training Center, Bowie, Maryland, and the Illinois Holocaust Museum and Education Center, Skokie, Illinois, are in the process of being certified as LEED Silver. The Pacific Garden Mission, Chicago, Illinois was successfully certified LEED Silver in 2009.

They believe that design is a process that is arrived at democratically through collaboration, so that the firm and its clients collectively evolve the work at hand rather than privileging that process for the designer alone. At a local level it speaks to interaction between client and architect, and at a larger level is a metaphor of those diverse strains that represent the unique American condition, defining their search for authenticity.

Index